Exploring Power Mechanics

Exploring power mechanics can lead to important, interesting discoveries. This "aeromobile," developed by a doctor in a small Illinois town, shows what one man can do. Powerful streams of air, forced down through the bottom of the machine, push it aloft and move it in any direction. This dramatic picture shows the vehicle completely off the ground as the doctor drives it through his workshop doorway in one of its early trials. Courtesy Peoria Journal Star.

In this engine the pistons lie flat, facing each other.

Courtesy Peoria Journal Star

EXPLORING POWER MECHANICS

By HAROLD T. GLENN

INSTRUCTOR
Public Schools
Long Beach,
California

AUTHOR
Automechanics
Safe Living
Youth at the Wheel

CHAS. A. BENNETT CO., INC.

Peoria, Illinois

PRINTED IN THE UNITED STATES OF AMERICA
Library of Cong. Cat. No. 62-7077

Fourth Printing 46VH510

CONTENTS

CHAPTER 1

POWER

To us, primitive man seems ignorant to have been a sun worshiper, but he was not far wrong in recognizing the sun as the most important thing in his life here—for without the sun he could not have existed. The sun is the source of energy for every kind of power used on earth, except nuclear. Human and animal muscles depend on food grown by the sun's energy. Wood, coal, and petroleum are the result of vegetation grown in the sun's rays. Wind and water power result from the weather, which is a direct product of the sun's heat.

The energy given off by the sun is generated by a nuclear reaction, similar to the "H" bomb. Hydrogen is constantly being changed into helium, which has a slightly smaller mass. Thus the sun grows smaller with each passing moment, the difference in mass being given off as energy. Einstein expressed the amount of energy available as a formula:

$$E = mc^2$$

E = energy
m = mass
c = velocity of light

The velocity of light is 186,000 miles per second, and when it is squared, the resulting multiplier (34,596,000,000) of even a very small amount of mass results in a tremendous amount of liberated energy.

One result of the sun's nuclear activity is a temperature in the millions of degrees. With all this energy being radiated, it is no wonder that man has always sought better ways of using it *directly*. Even before the birth of Christ, man learned to start fires by focusing the sun's rays through primitive concave metal mirrors. Several ways have been found to use the sun's rays directly to furnish power, but these have not as yet proved practical for large-scale applications. If man

The corona of the sun photographed during a total eclipse shows the enormous amount of energy being radiated about its surface.

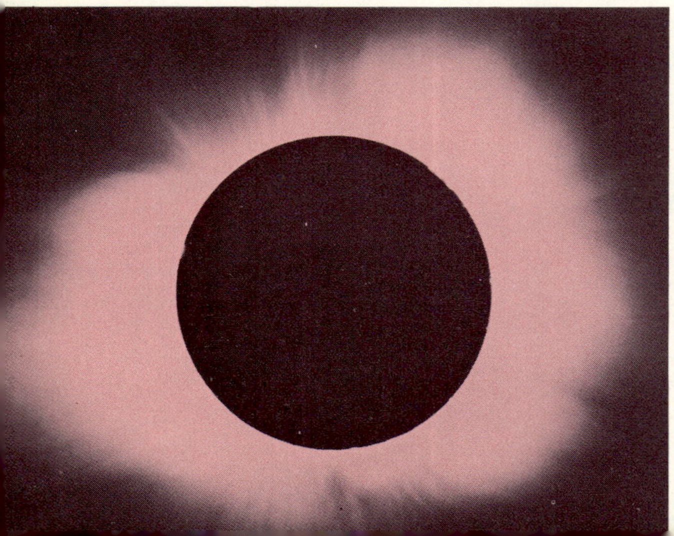

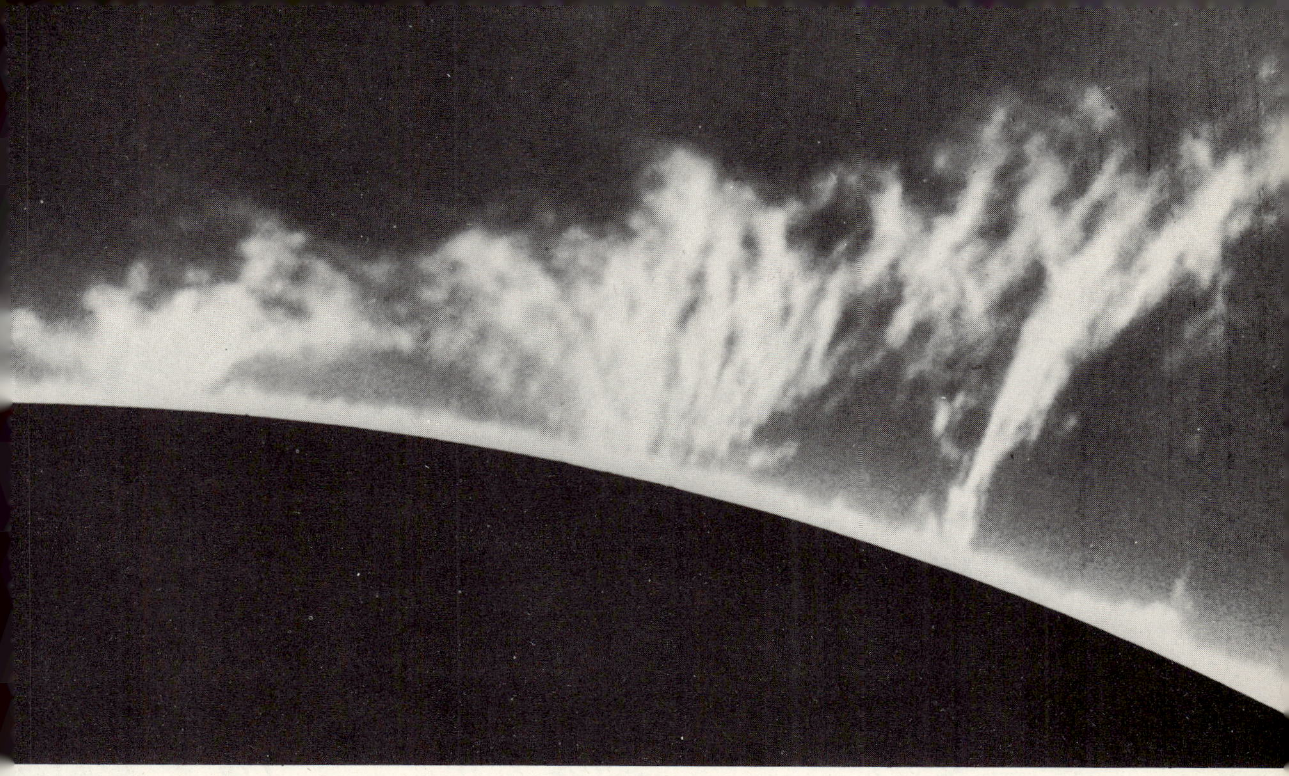

These displays appeared to reach a height of 80,000 miles.

could learn to get power directly from sunlight, the world would be changed drastically.

Most solar energy is used in an *indirect* manner, in the form of combustion—fire. When we burn a piece of wood, we oxidize carbon and hydrogen, a chemical process giving off heat. A typical chemical reaction looks like this:

$$CH_4 + 2\,O_2 = CO_2 + 2\,H_2O$$

CH_4 = methane (from petroleum)
O_2 = oxygen (from the air)
CO_2 = carbon dioxide
H_2O = water

About 1/10th of one per cent of the sun's energy which reaches the earth is stored in the wood when a plant or tree takes carbon dioxide from the air and water from the ground. These chemicals are changed into some form of carbohydrate and oxygen (photosynthesis), ready for burning. Writing it as a *simplified* equation, it looks like this:

$$CO_2 + H_2O + \text{energy} = \text{carbohydrate} + O_2$$

HISTORICAL DEVELOPMENT OF POWER

In early times, the only power available to man was his own muscle. No one has ever recorded the date when man first learned to use a tree limb as a lever to increase his ability to move a heavy object, or wheels to transport it from one place to another. But these simple devices revolutionized man's existence.

No doubt, it all began when some ancient man picked up a fallen tree limb, with a knurl or broken piece of branch on one end, and discovered that the long, weighted club increased the force of his blows. He didn't understand the principle of levers, but he understood that his hunting prowess became greater.

Today, we know that the weighted branch is a form of simple machine—a

Even ancient man learned how to use solar energy in the form of fire. He probably didn't know that the energy came from the sun but he did enjoy the warmth.

lever. Such a machine is a device to increase man's strength so that he can move heavy objects or strike harder blows. But note carefully that such machines *cannot increase man's ability to do work,* which is defined as "the product of the force and the distance through which it moves." Machines merely allow us to trade distance for "quick" force! For example, the wrench you use to loosen a tight nut is a form of lever. If the nut is stuck tightly, you may have to get a longer wrench to increase the leverage. Actually, you are not gaining in work accomplished, because you have to move the wrench handle a longer distance than the nut moves. You have to do the same amount of work, but in a different, *more gradual* form. You can either sacrifice speed to gain force, or sacrifice force to gain speed because, in every case, the work accomplished by a machine is always equal to the work applied. This important principle was first stated by Isaac Newton (England) in 1687.

SIMPLE MACHINES

It may interest you to learn that all of the complicated mechanism used in the operation of an automobile, its manufacture, and its repair combines six simple machines: the lever, wheel and axle, pulley, inclined plane, screw, and wedge. These are used to increase the force applied, change its direction, or change its speed. In the text which follows, each of these machines will be described and, in the rest of the book, there will be many references to the six simple machines, especially where illustrated, so that you can readily see their application.

Levers

A lever is a rigid bar free to move about a fixed point, called a *fulcrum.* There are three classes of levers: first, second, and third—according to the relative positions of the effort, resistance,

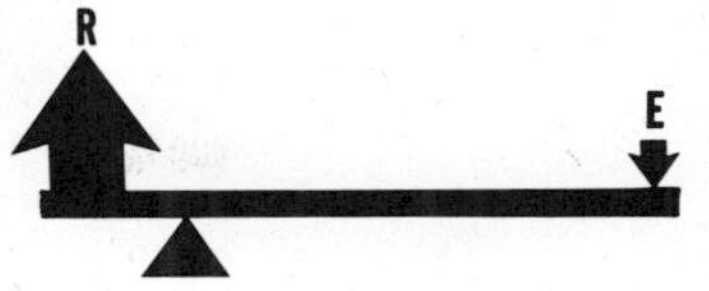

The first-class lever with the fulcrum placed between the effort and resistance is frequently used to change the direction of an applied force.

and the fulcrum. One great advantage of levers is that they operate with relatively little friction.

First Class Lever

Such a lever has the fulcrum placed *between the effort and the resistance.* If the fulcrum is placed closer to the

The first man who learned how to use a lever probably was known as a great hunter.

It was only a step from the club to a first-class lever. Indeed, "Necessity is the mother of invention."

resistance, speed is sacrificed for force, so that a heavy object can be moved by the application of a relatively small force. Or, if the fulcrum is placed closer to the effort, force is sacrificed for speed. In either case, *the work moves in a direction opposite to the applied force.*

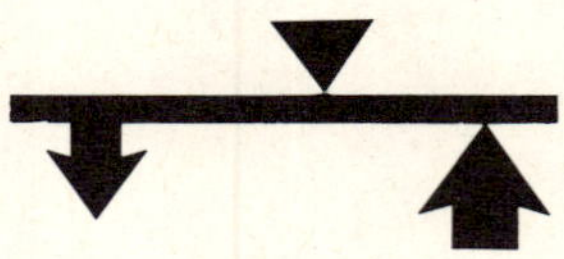

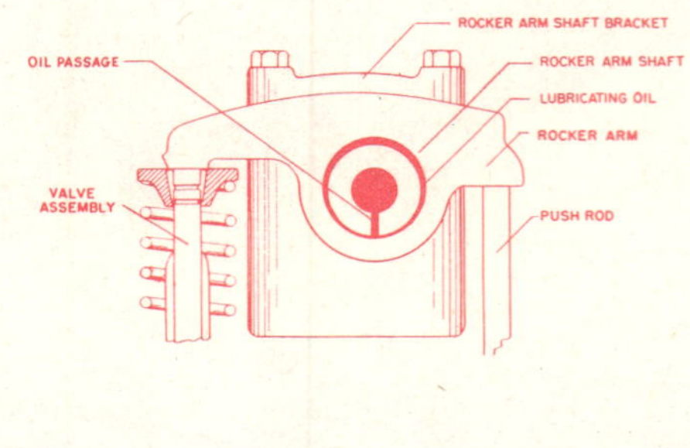

The first-class lever is used in many parts of an automobile. Here we see such a lever used in the form of a rocker arm to change the direction of motion. The push rod moves up and the valve moves down. Above the rocker arm is a diagram of the forces acting on the lever.

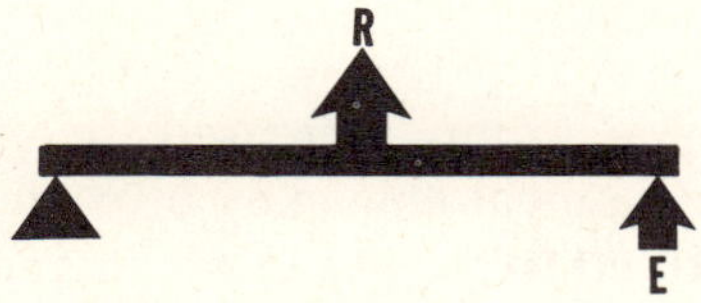

The second-class lever has the fulcrum at one end and the effort applied to the other.

Second Class Lever

This lever has the fulcrum placed at the end, with the resistance between the fulcrum and effort. Speed is sacrificed for force, and the direction of work movement is the same as the applied force.

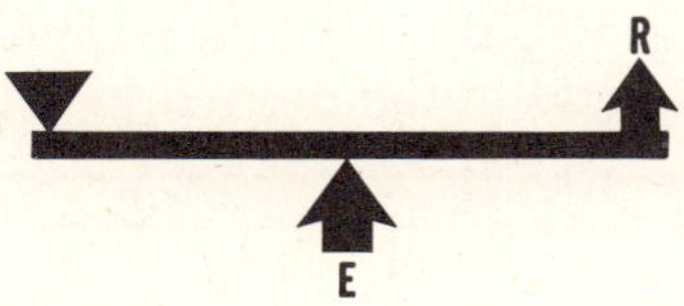

The third-class lever has the fulcrum at one end and the resistance at the other. A hammer is a good example of a third-class lever.

Third Class Lever

Such a lever is characterized by having the fulcrum placed at one end,

A wheel and axle create a form of lever. Note that you can simulate the application of leverage if you try to hold back the axle while the wheel is power driven.
Courtesy Barrett Equipment Co.

This valve spring compressor is a second-class lever. Courtesy Studebaker-Packard Corporation.

resistance at the other, and the effort applied to the middle. Force is sacrificed for speed, and the direction of work movement is the same as the applied force.

Wheel and Axle

A wheel and axle is a modified form of lever. A force applied to the rim of a wheel causes it, and its attached axle, to rotate. But, because the radii of both parts are unequal, the resulting force is greater if the resistance is placed on the axle. Conversely, if the effort is applied to the axle and the

This diagram of a lever imposed over the drawing of a wheel and axle shows the application of forces.

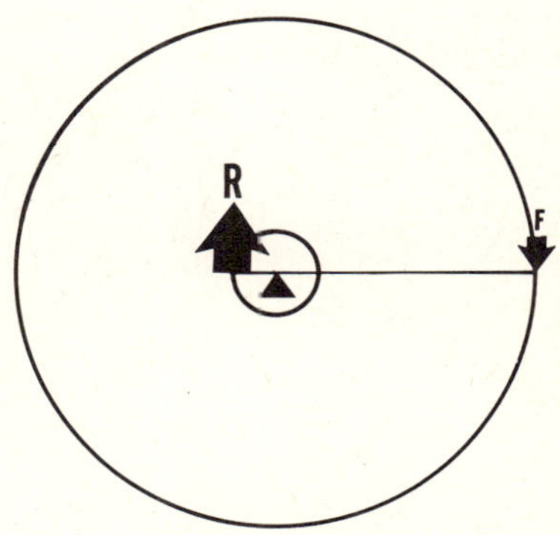

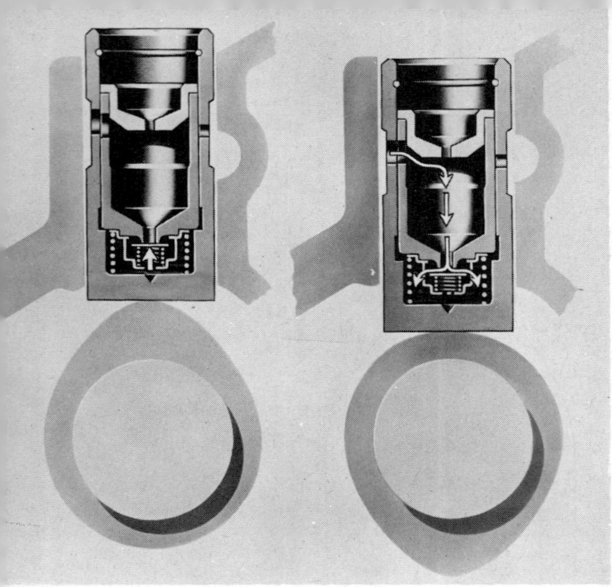

This camshaft operating a valve lifter is a good example of an inclined plane used in an automobile engine. Can you think of others?

resistance is placed on the rim of the wheel, the result will be a gain in speed and a decrease in force output.

Pulley

A pulley is a wheel that is free to rotate about an axle, in connection with a weight. In operation, it is similar to the wheel and axle just discussed and, therefore, is a form of lever. It has an advantage in that several pulleys can be combined in a block-and-tackle arrangement for increased leverage. The mechanical efficiency of each pulley is added to the others for increasing force.

An inclined plane is used to roll or slide an object from one level to another. If the angle is made to be 30° as shown in the drawing, what geometric theorem tells us that the length of the sloping surface is twice that of the height?

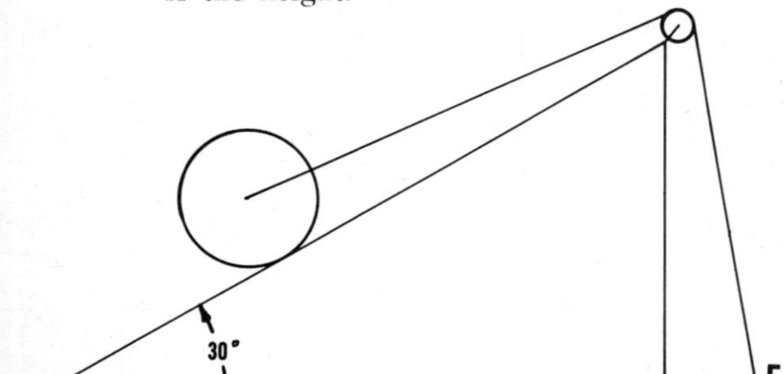

Inclined Plane

An inclined plane is a surface with a sloping side at an angle of less than 90° to the base. It is used to slide or roll objects from one level to another. For example, if the length of the sloping surface is made twice that of the height, the force required to move an object up will be half that required to lift it directly to the desired height.

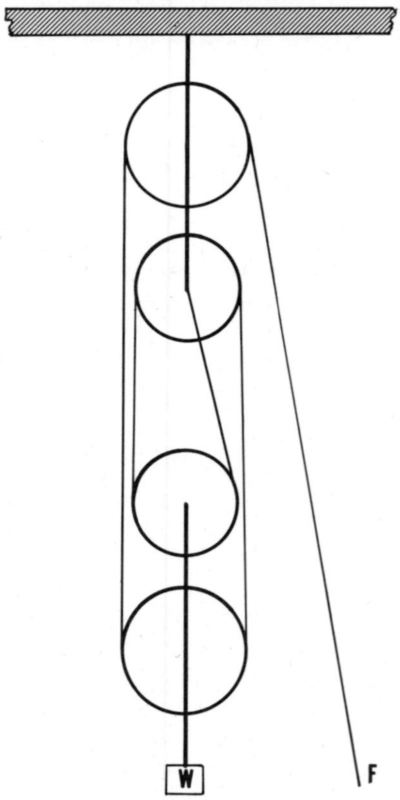

The pulley is a form of lever. It has the advantage that several pulleys can be combined into a block-and-tackle arrangement for increased leverage.

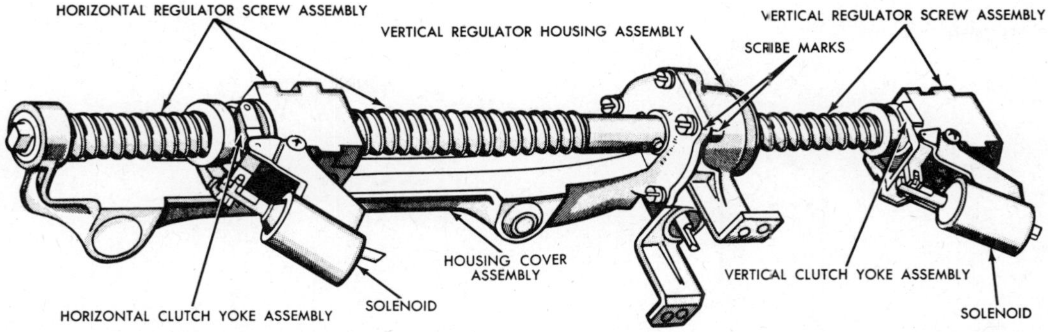

This screw assembly is used to move power-driven seats. Courtesy Ford Motor Company.

Screw

A screw is a modified inclined plane. The slope of the threads provides the movement when the screw is turned. A screw has high mechanical efficiency, but, unfortunately, also has very high frictional losses. By combining the high mechanical efficiency of the screw with the low frictional losses of ball bearings, engineers developed the illustrated mechanism, which is used to move power-driven seats.

Wedge

A wedge consists of two inclined planes, combined base to base. It is generally used in brake mechanisms, where movement of the wedge pushes brake shoes apart and into contact with the drum.

A wedge ("W") is used to force the brake shoes apart and into firm contact with the rotating brake drum. Courtesy Barrett Equipment Co.

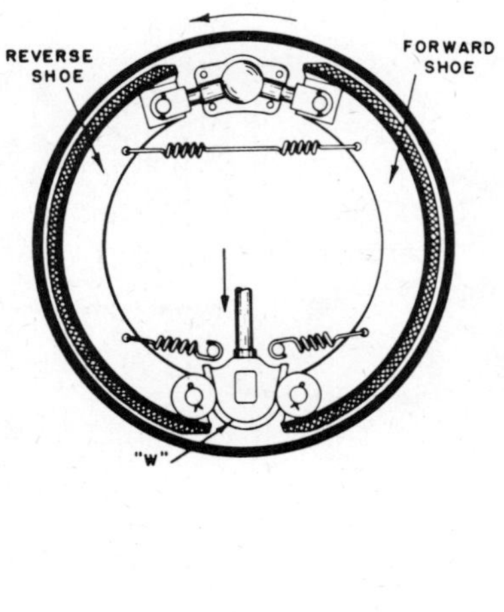

POWER-DRIVEN MECHANISMS

It is only within the past century that man has learned to use the six simple machines in complicated power mechanisms designed to do specific tasks. It was in 1705 that Thomas Newcomen (England) built an engine in which steam was condensed to produce a vacuum which drew a piston into a cylinder. It was crude, but it worked. Then James Watt (England) improved Newcomen's engine in 1769, greatly increasing its efficiency and power. After 1800, steam power progressed rapidly. A hundred years ago, about 25% of the work was done by machines; today the figure is near 95%.

Here the screw is used to remove a bearing assembly from a driveshaft. Courtesy Oldsmobile Division, General Motors Corporation.

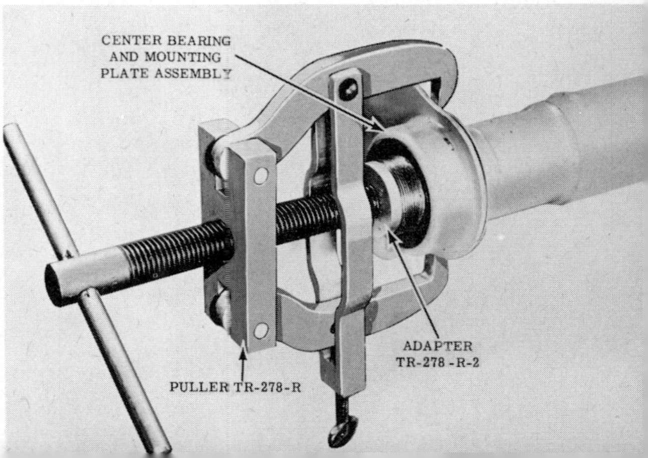

This is a diagram of Newcomen's steam engine. It operated on the vacuum developed in the cylinder when steam was condensed. Courtesy General Motors Corporation.

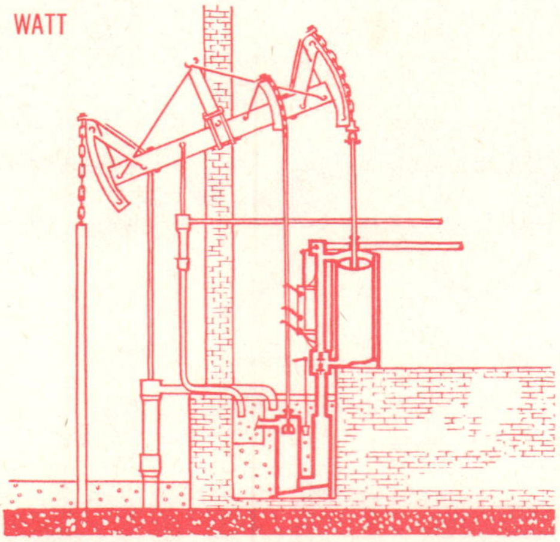

James Watt developed an engine which used the pressure of expanding steam to drive the piston into a cylinder. Courtesy General Motors Corporation.

Types of Engines

Traditional engines are of two basic types: *piston* or *turbine*. Either an expanding pressure is exerted against a piston contained within a cylinder, or a jet of pressure is directed against the blades of a turbine. The steam, gasoline, and diesel engines are of the piston type, while the gas and steam turbine engines are examples of the latter. The turbine engine is sometimes referred to as a reaction-type engine because it derives its power indirectly. The jet of steam is deflected by the turbine blades, which move because of reaction to the force.

The steam turbine revolves when steam pressure is directed against curved blades. A set of stationary blades, positioned between each pair of revolving blades, turns the steam and directs it against the face of the rotating blade assembly. Courtesy General Motors Corporation.

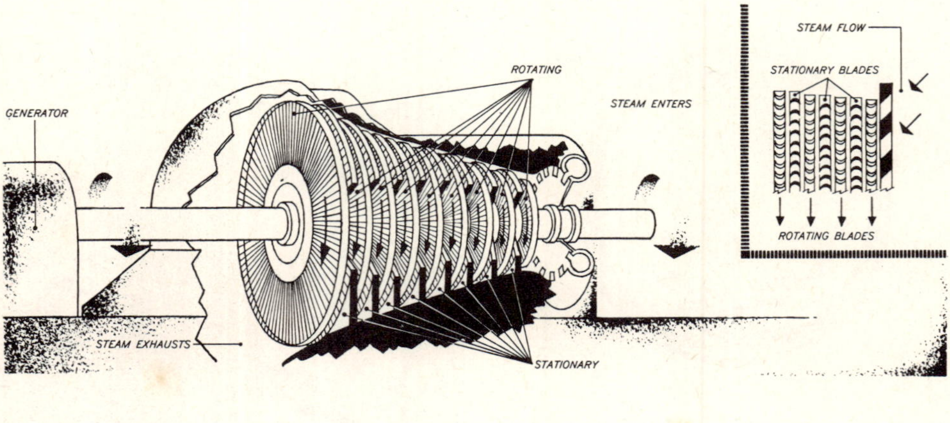

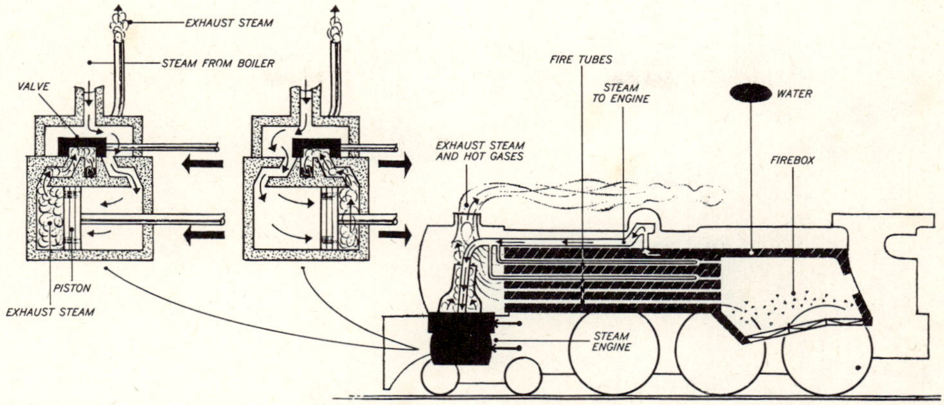

A cutaway drawing of a piston-type steam engine. The operation is explained in the text. Courtesy General Motors Corporation.

Steam Engine

While Watt's steam engine was originally developed for pumping water from mines, it soon found widespread use in the field of transportation. Although bulky, this source of power could be moved about. In 1769 Nicolas Joseph Cugnot (France) built the first self-propelled steam-powered gun tractor. It was heavy, under-powered, and soon abandoned, but it was the first vehicle to travel under its own power. After 1800 several successful steam locomotives were constructed which revolutionized travel—and so changed the lives of millions of people. For the next century, the piston-type steam locomotive dominated the scene.

In a piston-type steam engine, the steam is forced into one end of the cylinder, and then to the other, moving the piston back and forth. A sliding valve directs the steam into each end of the cylinder, in time with piston movement. When the valve is positioned to the left, steam enters the right end of the power cylinder while the left end is opened for exhaust. The expanding steam pushes the piston to the left, which forces the spent steam out through the exhaust port. Then the valve moves to the right, which opens the left end of the cylinder, forcing the piston to the right and starting the power cycle again.

Steam Turbine

Toward the end of the 19th century, a steam turbine-type engine began to replace the familiar piston-type engine

Joseph Cugnot's steam-powered gun carriage, built in 1769. Courtesy General Motors Corporation.

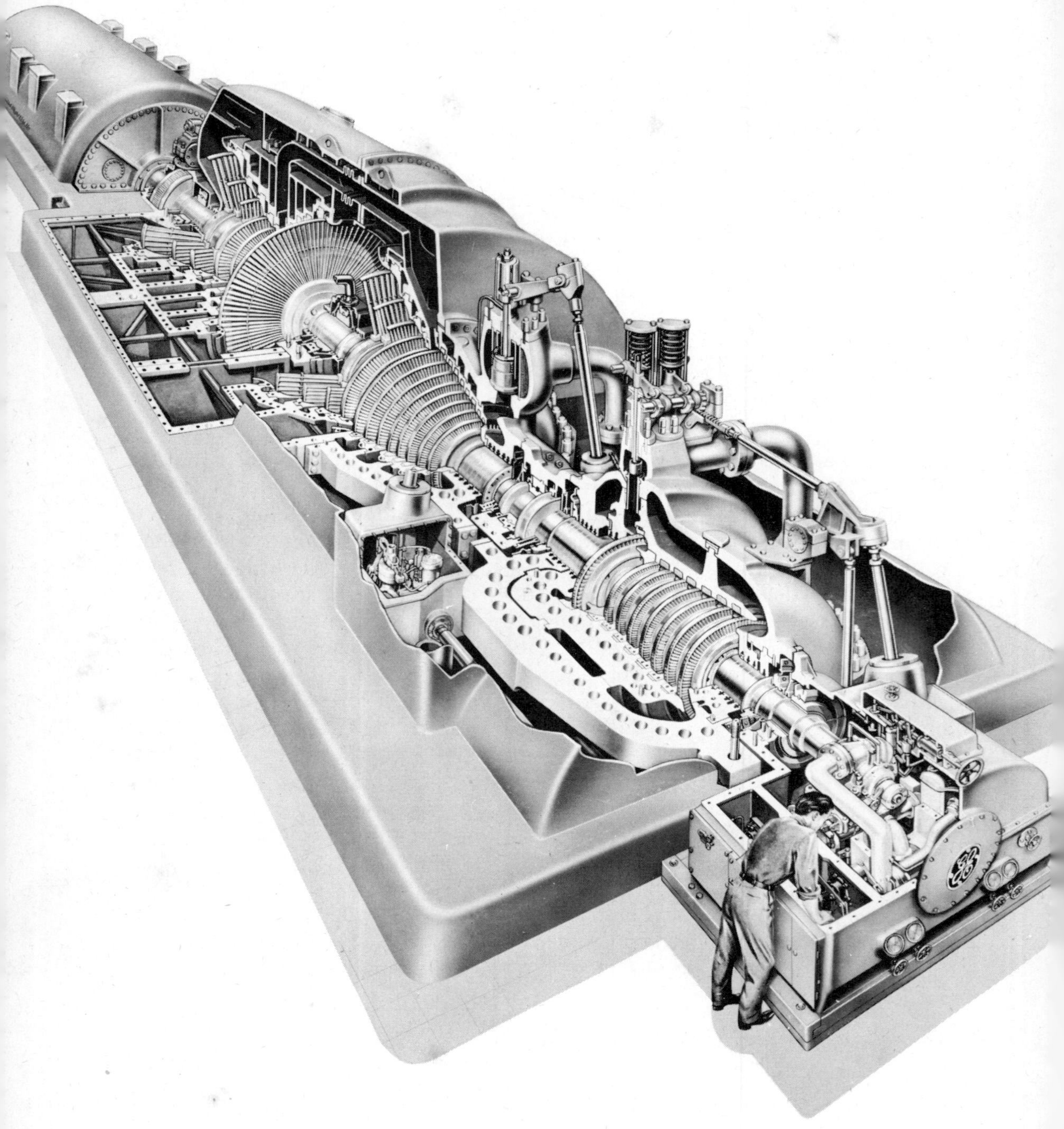

A modern steam turbine. Note its huge size in comparison to the size of the man in the foreground. The machine changes the energy of combustion (heat) into rotating motion, which produces electricity by friction of magnetic lines of force (induction) in the generator at the rear. Courtesy General Electric Company.

This shows the curvature of the vanes of both the stator and compressor blades. Note how the stages develop as the air moves to the left end of the machine. Courtesy Allison Division, General Motors Corporation.

When an atom of uranium 235 is struck by a neutron, it splits into barium and krypton. Courtesy General Motors Corporation.

for large installations. The turbine is much smoother in operation because it has no reciprocating parts (back and forth). The expanding steam pushes against the blades of a revolving shaft to provide the motive power. Both Charles G. Curtis (America) and Sir Charles Parsons (England) perfected the turbine by fastening a number of turbine wheels to one shaft so that the steam energy could be put to work several times before being exhausted.

In the steam turbine engine, a boiler (external combustion) produces steam, which is directed against a series of curved blades fastened to a revolving shaft. Then the steam jet direction is changed by a set of reactor blades, fastened to the stationary housing which surrounds the rotating turbine. The direction is changed so that it can push against the next set of movable blades. Thus the steam follows a zigzag path as it travels from stage to stage through the length of the turbine. As it is constantly expanding, the turbine wheels are increased in size to make use of the remaining energy.

Nuclear Power

The newest form of power is heat developed by a nuclear reaction. When uranium (U 235) is bombarded with neutrons, it splits (fission) into barium (Ba 137) and krypton (Kr 83) which, when added together, have slightly less mass than the original uranium. The difference is given off as energy according to Einstein's equation $E = mc^2$. Because the speed of light squared is so large, it takes only a very small amount of matter to produce huge quantities of energy. For example, the changing of one ounce of uranium into energy is equivalent to the energy output of the Hoover Dam hydroelectric power plant for an entire month.

A nuclear reactor contains fissionable material separated by graphite so that the action can be controlled. The heat of reaction generates steam, which drives a turbine in the usual manner. In other words, a reactor furnace is "substituted" for an oil- or coal-fired furnace.

The heat of fission is so intense that liquid sodium must be pumped through the reactor to extract it. The sodium is then passed into a heat exchanger where the energy is transferred to a water-filled system to be turned into steam. The sodium system must be shielded because it becomes radioactive from passing through the reactor and is, therefore, unsafe to man.

Piston-Type Engine

A piston-type engine operates similarly to a cannon. If you think of the barrel as the cylinder, the cannon ball as the piston, and the gunpowder as the fuel, you have the elements of a piston-type engine. Igniting the gunpowder causes gases to form which expand and force the cannon ball from the barrel.

In 1680 Christian Huyghens (Holland) described such an engine. However, his piston responded to the vacuum developed after the combustion gases were cooled. In 1876 Dr. Nicholas August Otto (Germany) built an engine in which expanding gases pushed a piston down a cylinder. In addition, he added two valves to allow a fresh fuel charge to enter, and the exhaust gases to leave the cylinder. His engine worked on the four-stroke cycle principle—one which we use today and which retains his name—the Otto cycle.

An atomic energy engine is basically a steam turbine. Liquid sodium is used to remove the heat of fission, transferring it to a fluid coolant system in the heat exchanger.

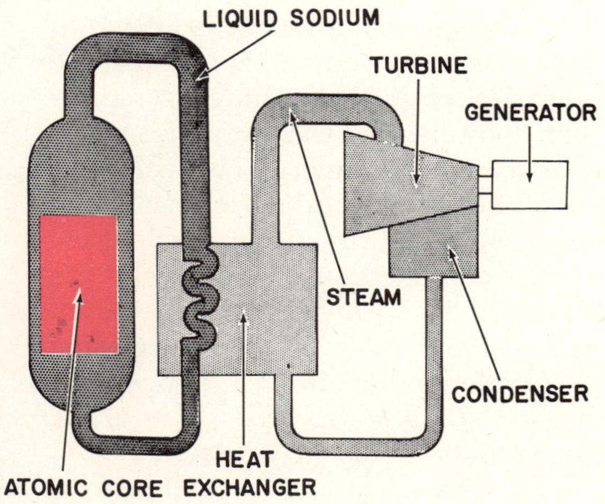

Four-Stroke-Cycle Gasoline Engine (Otto Cycle)

The piston, which moves up and down within a cylinder, is connected to a crankshaft which, in an automobile, is coupled to the rear wheels through the transmission and rear drive mechanism. The crankshaft and the connecting rods change the up-and-down motion of the piston into a rotary one.

Intake Stroke

As the piston moves down, the re-

The four strokes of a four-stroke-cycle gasoline engine.

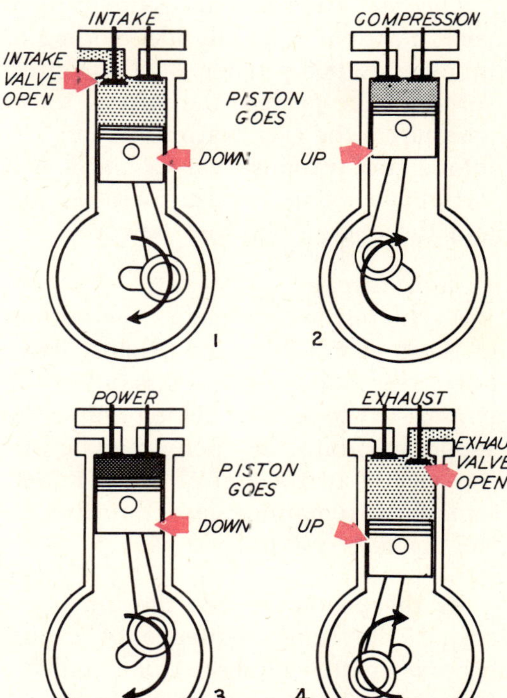

sulting "suction" draws the fuel-air mixture through the carburetor, through the open intake valve, and into the combustion chamber. At the bottom of the down stroke, the intake valve closes and seals the combustion chamber, making it airtight.

Compression Stroke

Next, the piston moves up in the cylinder, compressing the flammable mixture.

Ignition

When the piston is close to the top of the stroke, the spark ignites the fuel particles close to the spark plug which starts the mixture burning. The flame passes quickly, in about 1/350th of a second, through the compressed fuel-air mixture.

Power Stroke

The resulting heat causes the trapped gases to expand rapidly, developing as much as 12,000 pounds, or 6 tons pressure, which pushes the piston down and moves the car. Naturally, both the intake and exhaust valves must seat airtight, or some pressure will escape, and the engine will lose power.

Exhaust Stroke

As the piston approaches the bottom of the power stroke, the exhaust valve opens, the burned gases pass out of the engine, to the muffler, which quiets the sound, and into the open air. The piston returns to the top of the cylinder, forcing the remaining gases out through the exhaust system.

As the piston reaches the top of its stroke, the same sequence of events occurs again: intake stroke (piston goes down—intake valve open—fuel-air mixture is drawn into the cylinder);

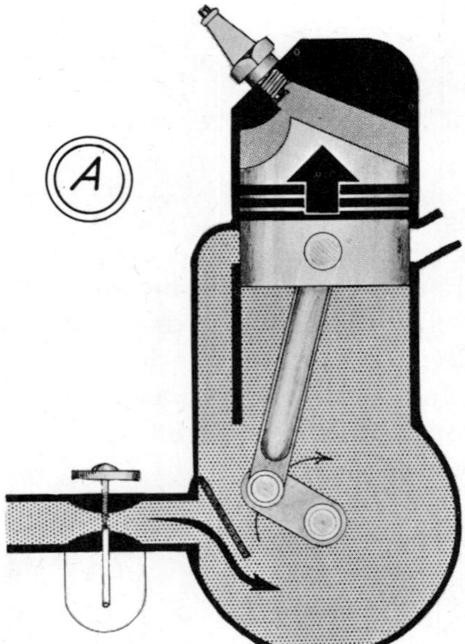

The intake stroke of a two-stroke-cycle gasoline engine.

compression stroke (piston goes up—both valves closed—mixture is trapped and compressed); ignition (spark passes through the fuel-air mixture, which burns rapidly); power stroke (piston driven down — both valves closed); and exhaust stroke (piston goes up—exhaust valve open—burned gases forced from the cylinder).

This sequence of events is known as the four-stroke cycle because of the four strokes involved: intake, compression, power, and exhaust. This cycle is repeated continually. Two complete revolutions are necessary for each power stroke to occur, or for each cycle to be completed once. The cycles are arranged in such an order that each cylinder is fired in correct succession to develop a smooth flow of power.

Two-Stroke-Cycle Gasoline Engine

The two-stroke cycle engine fires

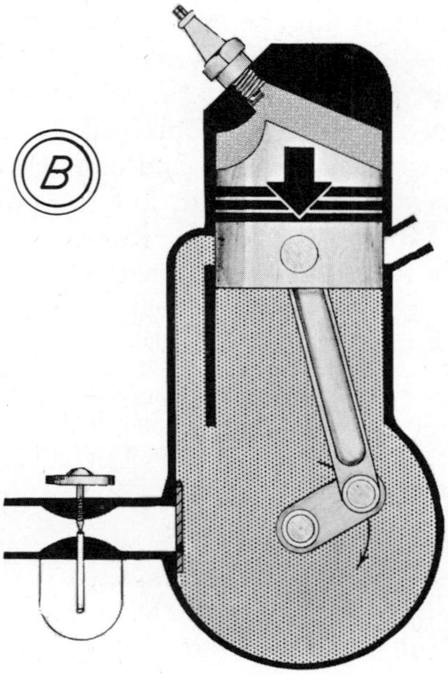

The compression stage of the fuel charge in the crankcase.

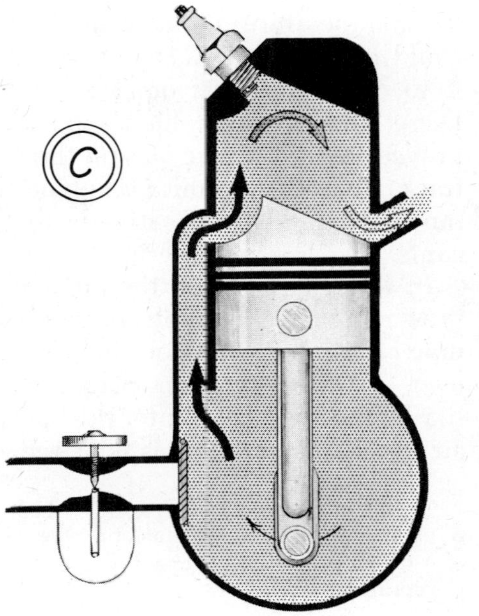

The passing of the fuel charge into the combustion chamber, driving out the remaining exhaust gases.

every time the piston reaches the top of the cylinder. As in a four-stroke cycle engine, the carburetor mixes the fuel and air which is drawn into the crankcase through a reed valve by suction as the piston moves to the top of the cylinder.

As the piston moves down, the fuel charge in the crankcase is compressed, which closes the reed valve and prevents movement of the fuel-air mixture back through the carburetor. When the intake port is uncovered, the compressed fuel-air mixture in the crankcase is forced into the combustion chamber, pushing the exhaust gases through the exhaust port.

As the crankshaft continues to rotate, the piston moves up, sealing off the intake and exhaust ports and compressing the fuel charge. When the piston reaches the top of its stroke, the spark plug fires and ignites the mixture which expands and drives the piston down.

When the piston moves down far enough to uncover the exhaust port, the exhaust gases start to move from the cylinder. As the piston moves farther down the cylinder, the intake port is uncovered and a fresh charge enters the combustion area, forcing out the remaining exhaust gas. Each time the piston reaches the top of its stroke, a charge is burned.

Diesel Engine

About the same time the internal combustion gasoline engine was being developed, Dr. Rudolph Diesel (Germany) experimented with an engine in which fuel was ignited by the heat of compression. He based his work on the experiments of James Prescott Joule (England) who discovered that the temperature of a confined gas could be raised 87°C. by compressing it to one half of its original volume. Diesel's first engine blew up, which proved to him that he was on the right track. In 1877 he built an engine that ran successfully and still bears his name.

In the diesel engine the air is compressed to 1/16th of its original volume, raising its temperature to well over 1,000°F. Ignition occurs when the fuel is injected into this heated air so that a spark plug is not needed.

A cross-section of a fuel injector, the heart of a diesel engine. Courtesy General Motors Corporation.

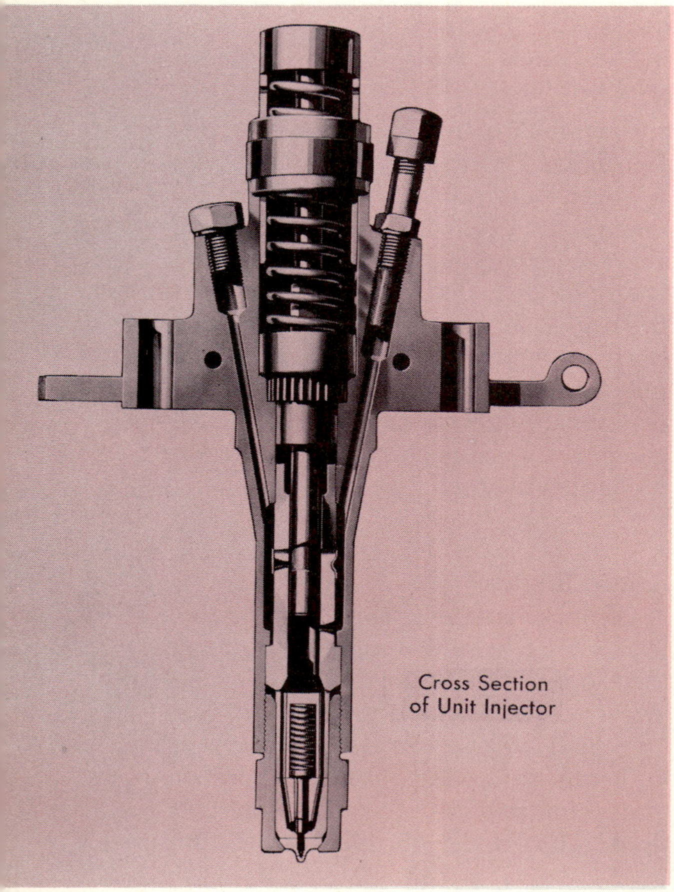

The first diesel engine worked on the four-stroke cycle principle. A comparison of the strokes of the gasoline and diesel engines (four-stroke cycle) shows that the intake and compression strokes vary somewhat, but the power and exhaust strokes are the same:

Stroke	**Gasoline engine**	**Diesel engine**
Intake	Fuel-air mixture drawn into the cylinder	Air drawn into the cylinder. Fuel sprayed into cylinder at the end of compression stroke
Compression	10-1	18-1
Ignition	By spark	By compression
Power	The same	The same
Exhaust	The same	The same

Mechanically, there is very little difference between the two engines, both having cylinders, pistons, connecting rods, and a crankshaft. Both have mechanically operated poppet

This dramatic picture shows the spray pattern from the nozzle of a fuel injector. The fuel particles move at about 600 mph. Courtesy General Motors Corporation.

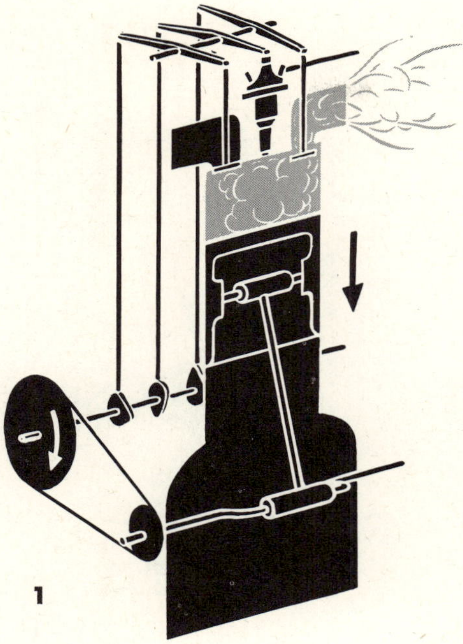

1

The intake stroke of a four-stroke-cycle diesel engine. Courtesy General Motors Corporation.

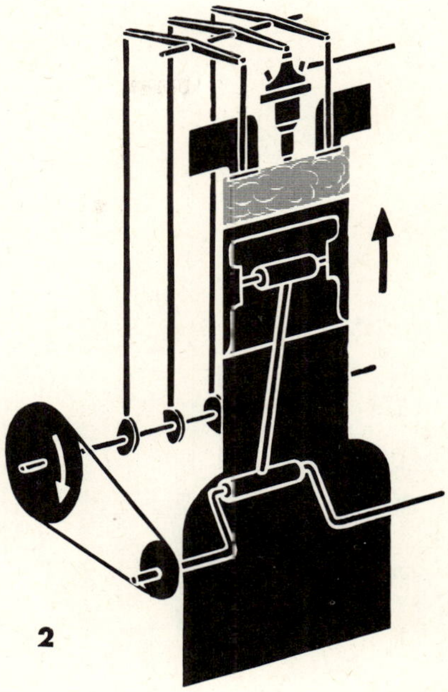

2

The compression stroke. Courtesy General Motors Corporation.

The power stroke. Courtesy General Motors Corporation.

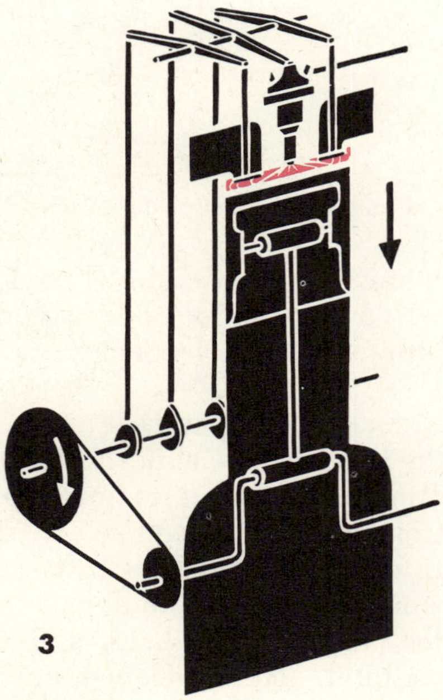

3

The exhaust stroke. Courtesy General Motors Corporation.

4

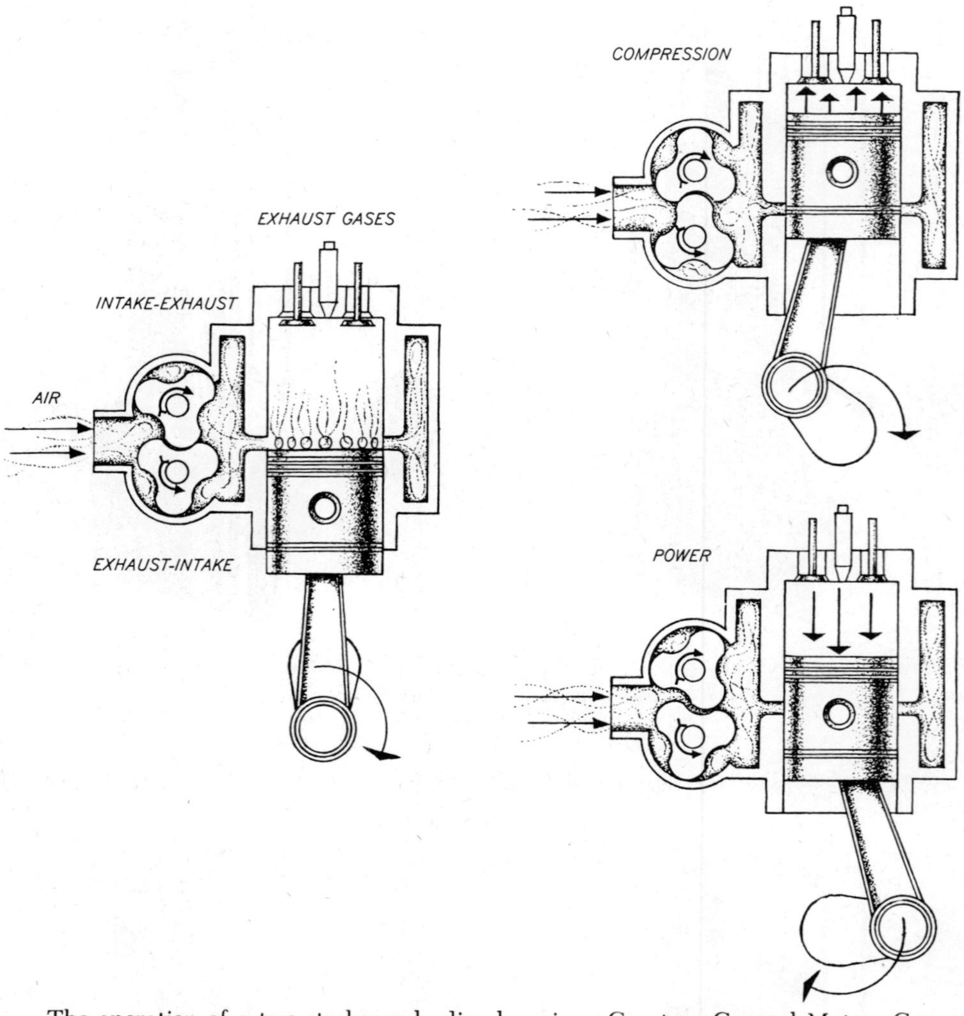

The operation of a two-stroke-cycle diesel engine. Courtesy General Motors Corporation.

valves, intake and exhaust manifolds, timing gears, a camshaft, and a flywheel. Both have lubricating and cooling systems, and in general, they look very much alike. Because of the greater operating pressures, the diesel engine parts are heavier.

In both engines, the output is controlled by varying the amount of fuel fed to the cylinders. In the gasoline engine, this is done by the *carburetor throttle,* which restricts the flow of fuel-air mixture. In the diesel engine, the fuel flow is controlled by varying the output of the *fuel injector.*

Two-Stroke-Cycle Diesel Engine

In the late 1920's, Charles F. Kettering (America) set out to develop a two-stroke cycle diesel engine. He felt that an engine which spent half of its time pumping air into and out of a cylinder (intake and exhaust strokes) was wasteful. He developed an air blower to take the place of both strokes. The pump forced fresh air into the cylinder which was used to push the burned gases out so that each up stroke could be used to compress air, ready for injection and a new power stroke. This resulted in a great saving in weight because such an engine developed almost twice the power of the older, four-stroke cycle engine.

When the piston approaches the bottom of its stroke, the exhaust valve opens to start the exhaust gases moving from the cylinder under the pressure remaining from the burning charge. As the piston moves farther down the cylinder, it uncovers ports into which the blower forces fresh air, which drives the remaining burned gases from the cylinder.

As the piston starts up, it covers the intake ports and compresses the air. At the top of the stroke, fuel is injected and ignition occurs, driving the piston down on the power stroke. When the piston approaches the bottom of the stroke, the exhaust valve opens, the intake ports are uncovered, and the cycle begins again. A power stroke occurs each time the piston reaches the top of its stroke.

Engine Power Computations

James Watt introduced the term horsepower in order to describe the ability of his engine to do useful work. He said that whenever a weight is moved, work is being done. And when the work is performed at the rate of 33,000 ft. lbs. per minute, one horsepower of energy is being expended which leads to the universally used formula:

$$\text{Hp} = \frac{\text{ft. lbs. of work per minute}}{33{,}000}$$

S.A.E. RATING

There are several ways in which engine power can be determined. It can be calculated by the following formula:

$$\text{Hp} = \frac{D^2N}{2.5}$$

D = bore of cylinder in inches
N = number of cylinders
2.5 = constant

For example, if an eight-cylinder

Sixty years of progress from horse power to hp. Courtesy Socony Mobil Oil Company.

engine has a bore of three inches, the calculations are as follows:

$$Hp = \frac{3^2 (8)}{2.5} = \frac{9\ (8)}{2.5} = \frac{72}{2.5} = 28.8 \text{ hp}$$

This formula is generally used to calculate the taxable hp of an engine as it results in a lower figure than the developed hp, which means a lower tax.

Brake Horsepower

The power output of an engine can be accurately determined by changing its energy into electricity or heat. In the first instance, the engine is connected to an electric generator, the output of which is measured with gauges. A Prony brake is used to change the engine's rotating energy into heat. It consists of a lined band which can be tightened around the flywheel of an engine until the speed is slowed to the desired rpm. The resulting torque is measured on a scale and the bhp calculated by the formula:

$$Bhp = \frac{F \times 2L \times 3.1416 \times rpm}{33{,}000}$$

Substituting a set of values for a specific engine running at 2,000 rpm, where the scale reading is 100 lbs., and the length of the lever arm is 3 feet:

$$Bhp = \frac{100 \times 2(3) \times 3.1416 \times 2{,}000}{33{,}000}$$

$$= \frac{3{,}769{,}920}{33{,}000} = 114.27 \text{ bhp}$$

Engineers are testing the hp output of an engine by changing its energy into electricity, which can be measured on these gauges. Courtesy General Motors Corporation.

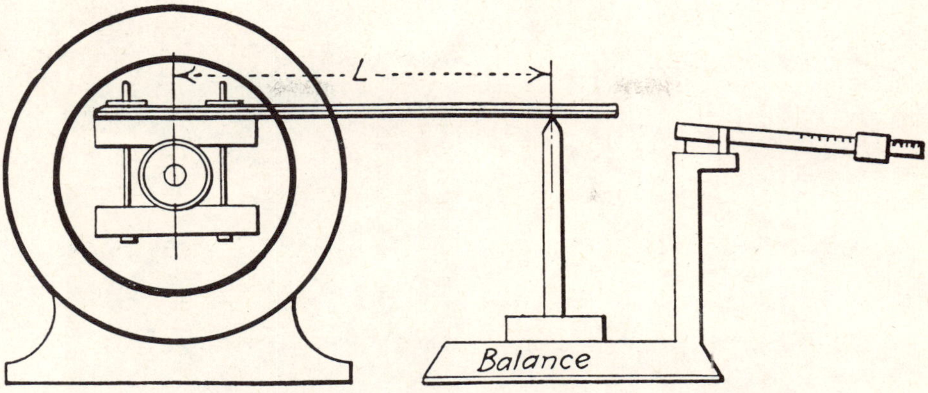

The Prony brake is used to determine the torque of a motor.

Another form of Prony brake consists of two spring scales T_1 and T_2 which can be adjusted to load the engine in a similar manner. The difference in scale readings is substituted in the following formula to calculate the bhp:

$$\text{Bhp} = \frac{(T_2 - T_1) \times d \times 3.1416 \times \text{rpm}}{33{,}000}$$

Computing the brake horsepower of an engine running at 2,000 rpm, where the difference in scale readings is 200 lbs. and the diameter (d) of the flywheel is 2 feet:

$$\text{Bhp} = \frac{200 \times 2 \times 3.1416 \times 2{,}000}{33{,}000}$$

$$= \frac{2{,}513{,}280}{33{,}000} = 76.16 \text{ hp}$$

The spring balance method can also be used as a Prony brake.

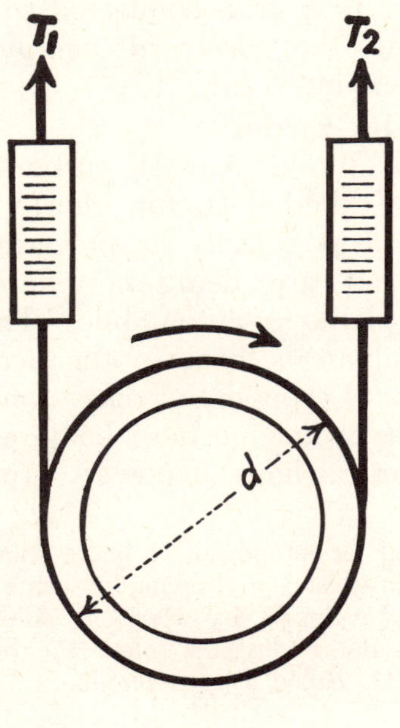

Electric Power Computations

The amount of power used by an electric motor is a direct indication of its power developing capabilities. The

A typical electric motor test to determine the output of a cranking motor. Courtesy Delco-Remy Division, General Motors Corporation.

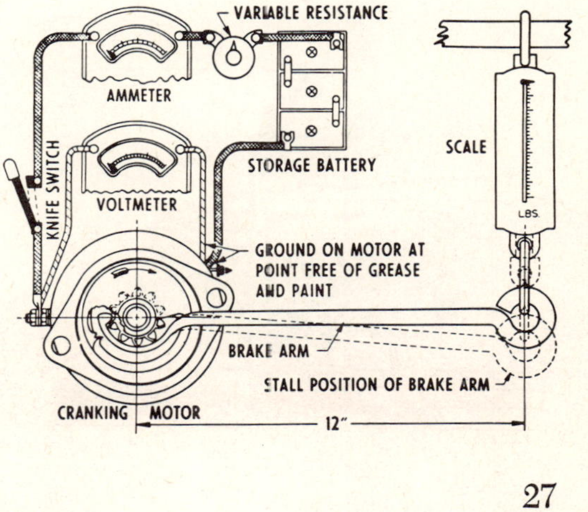

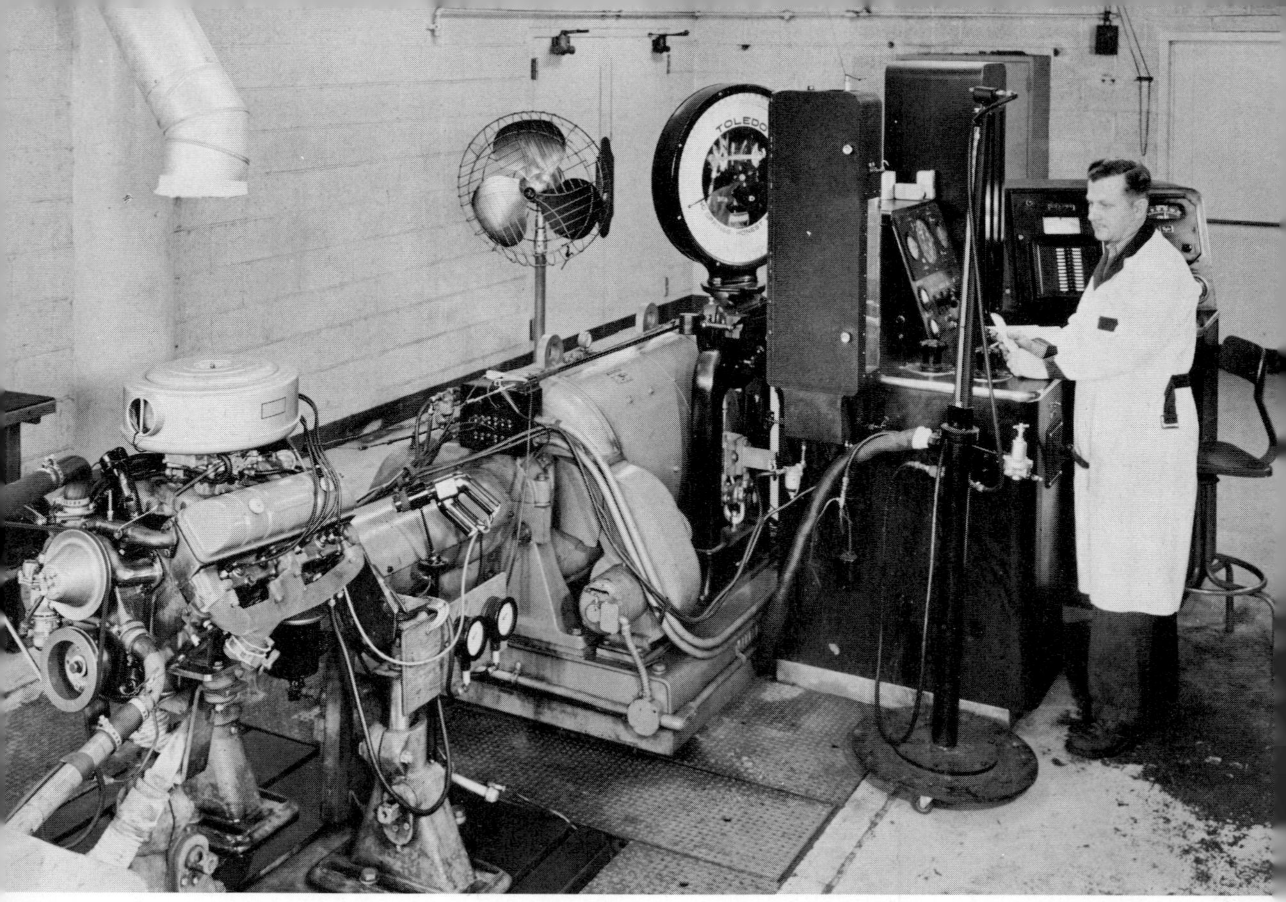

This engineer is determining the hp of an engine. Note the scale in the background to measure the torque. Courtesy American Motors Corporation.

power (Watt) is the product of the amperage the motor draws and the voltage under which it operates. It has been established that an electric motor which consumes 746 watts is developing one horsepower.

By coupling a gasoline engine to a generator, engineers can calculate its output by noting how much power it supplies to a grid, connected so as to consume the electrical energy and change it into heat.

Turbo-Jet Engines

Since World War II, turbo-jet engines (called jets for short) have gained in popularity for powering aircraft and are being considered as power plants for automobiles. This has come about because of the need for speed and economy of operation. The piston-type engine has about reached its practical limit of power output for

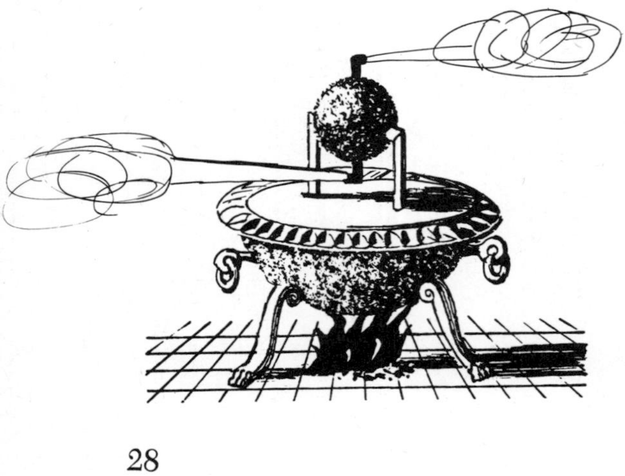

We think of jet power as being something new, but this is a drawing of a steam jet engine developed by Hero of Alexandria about a hundred years before the birth of Christ. He called it an Aeolipile.

Sir Isaac Newton designed this jet-propelled carriage in the 17th century. Courtesy General Motors Corporation.

aircraft because increasing it further would mean intolerable extensions in size, weight, and complexity.

Jet engines pack enormous power into a smaller space than piston-type engines, therefore they offer less air resistance to the forward motion of the plane. They have fewer moving parts and are insensitive to fuel requirements. Because of the lower air density at high altitudes, a jet plane operates more efficiently than a propeller-driven plane which runs out of air to bite into.

An inflated toy balloon can be used to demonstrate the principle of a jet engine by releasing the stem. The air rushes out, and the balloon darts off seemingly in every direction at once. Actually, it is always traveling in a

The turbo-jet lessens air resistance and has fewer moving parts than the piston-type propelled engine. Courtesy General Motors Corporation.

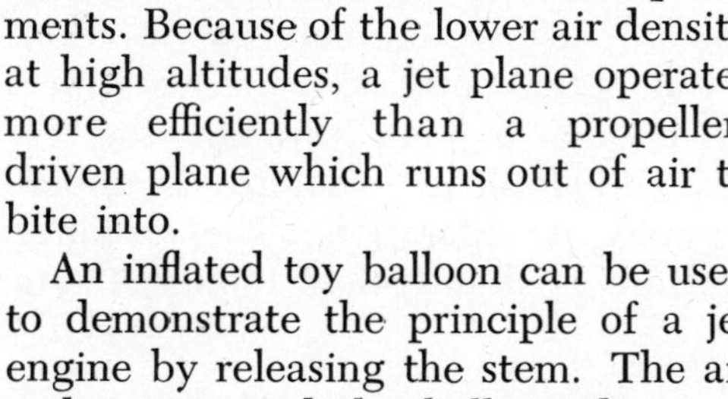

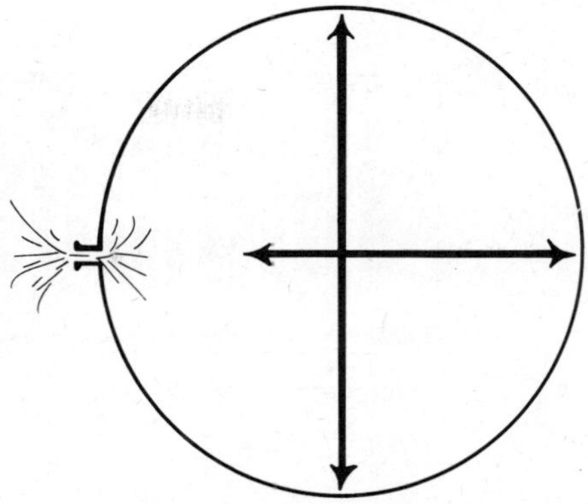

A balloon can be used to demonstrate the principle of a jet engine. The arrows represent the forces involved (vectors); each vector's length represents the strength of the force and the arrowhead shows its direction. Note that the opening at the left allows pressure to leak out, so this vector is shorter than the others—unbalancing the forces involved. The balloon moves to the right because that vector is stronger than the one to the left.

direction opposite to the jet of air blowing from the stem. When the stem is held tightly closed, the air pushes against all sides of the balloon equally, and no movement occurs. When the stem is released, the pressure is lowered on the stem side so the balloon travels away from the stem. Movement continues as long as there is pressure inside of the balloon.

In a jet engine, air enters the front, is compressed, mixed with fuel, and ignited. The resulting expansion forces a continuous stream of hot gases from the back of the engine at very high speed. The *reaction* to the large force necessary to push the gas from the engine is the force which pushes the plane forward.

All types of aircraft power plants produce their propulsive force by accelerating a mass of air. In a propeller-driven plane, a proportionately

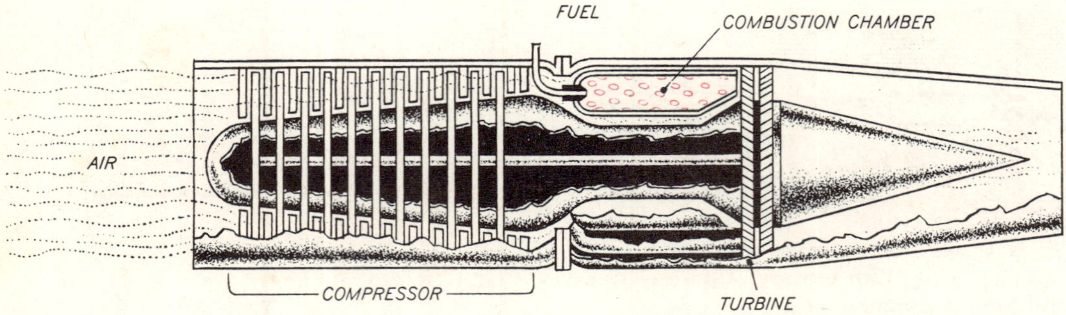

The basic functioning of a jet engine is simplicity itself compared to a piston-type engine. Air enters from the left, is compressed, expanded by combustion, and expelled at the rear. Courtesy General Motors Corporation.

large amount of air is given a relatively small increase in velocity; with turbojets and ramjet engines, a smaller amount of air is given a larger acceleration. Jet engines operate on the principle based on Newton's third law of motion: "For every action, there is an equal and opposite reaction." This is why jet engines are sometimes called reaction-type engines.

Unlike a piston-type engine, a jet operates with a continuous flow of power. Fuel and air are constantly being forced into the engine and combustion goes on continuously. Also, because all moving parts of the engine rotate instead of reciprocating (moving back and forth as in a piston-type engine), a jet operates with much less vibration.

The actual conditions that exist in a jet engine moving through the air about 450 mph are of interest in understanding its operation. The entrance pressure is slightly less than the ram

These are the actual conditions that might exist in a jet engine. The propulsive force is caused by the difference in the speed of the air as it enters the front of the engine and then is expelled from the rear. Courtesy General Electric Company.

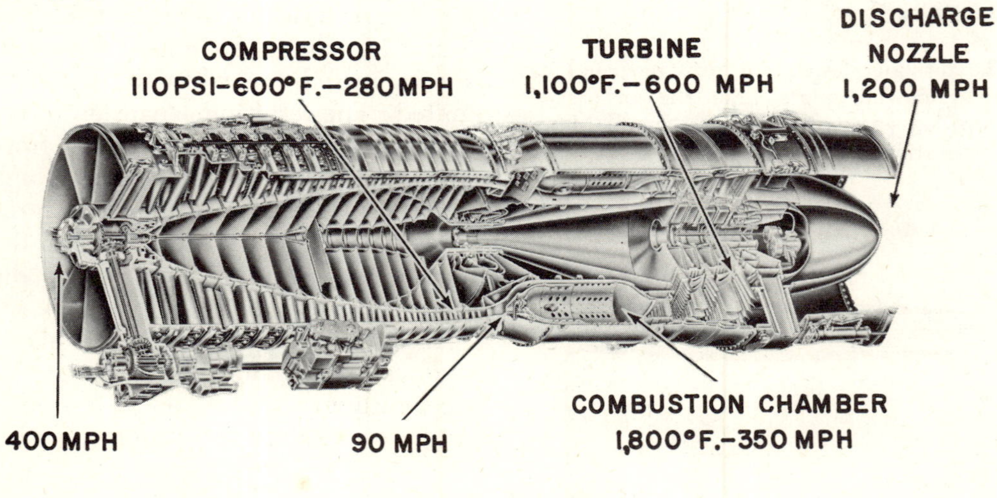

A close-up view of the combustion chamber and other details. Note the holes in the walls to allow a cooling curtain of air to enter, keeping flames from directly contacting the metal surface. Courtesy Pratt & Whitney Aircraft Company.

pressure because of the compressor action, being around 400 mph. As the air moves through successive stages of the compressor, its temperature and pressure increase while its velocity decreases. With a 9 to 1 compression ratio, the exit pressure is about 110 psi, the temperature close to 600°F., and the velocity about 280 mph. The air divides into two streams; one (with about 25 per cent) is slowed to about 90 mph, enters the combustion chambers, mixes with fuel, and ignites. The secondary air flows through holes in the combustion chamber walls to act as a coolant curtain which keeps the flames from striking the walls. Combustion adds energy to the air stream, increasing its temperature to about 1,800°F. and the velocity to about 350 mph. The gas stream passes through the turbine wheel, which removes energy as it drives the compressor so that the temperature drops to about 1,100°F. But, because of the convergent nature of the blading and vanes, the velocity increases to about 600 mph. By passing the exhaust gases through a convergent duct at the rear of the engine, the velocity of the jet stream is further raised to about 1,200 mph. The *difference* between the speed of the gases *leaving* the discharge nozzle and of the air *entering* the front of the engine is a measure of the thrust developed by the engine. This follows

This dramatic view shows a jet engine operating on a test stand. Courtesy General Electric Company.

the law of physics which states that the force is equal to the mass times the velocity:

$$F = Ma$$

FUEL SYSTEM

The *fuel control* is the "carburetor" of a jet engine. It meters the fuel according to the speed, altitude, and temperature and forces it through nozzles from where it is sprayed into the fiercely burning gas.

Some jet engines are equipped with an *afterburner* which sprays fuel into the exhaust stream after it passes the turbine. This expands and speeds up the gases. Because this combustion is behind the turbine, it does not add heat or pressure to the already overloaded turbine blades. An afterburner can increase engine power almost ⅓, but in doing so it consumes unusual amounts of fuel. Therefore, it is used only when emergency power is required.

The fuel used is kerosene, and since a jet engine is not sensitive to fuel requirements, a wide range can be used. One important specification is that the moisture content be held to a minimum because of the danger of icing at high altitudes where the temperature is often 80°F. *below* zero. Ice in a fuel line restricts the flow to the engine and a "flame-out" results.

IGNITION

The ignition system is needed only once, when starting the engine, after which the burning gas ignites the fresh charge. An ignition coil steps up the battery voltage similar to an automobile ignition system, but a vibrating interrupter replaces the usual contact points as no critical timing is needed. Many jet engines use a capacitor-

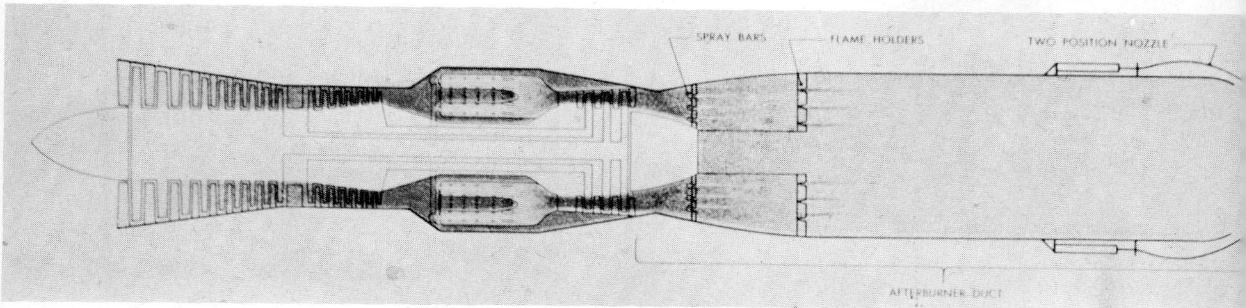

An afterburner is essentially a "stovepipe," fastened to the rear of a jet engine. It consumes as much air in five seconds as the average small home would hold. Courtesy Pratt & Whitney Aircraft Company.

discharge type of ignition system, containing a condenser to store up electrical energy which can be released at the proper time. Its main advantage is that it provides a much hotter spark, desirable when starting a dead engine at high altitudes or when the ground temperature is low.

An afterburner requires a separate ignition system because the fuel is injected quite a distance behind the combustion chambers. Two nozzles are located just back of the turbine which spray fuel into the burning gases. The fuel ignites and two streaks of flame are carried back to ignite the main fuel being squirted from the afterburner nozzles.

Turbo-Prop Engines

In a turbo-prop engine, the arrangement of parts is similar to the turbo-jet engine just discussed, except that a propeller is fitted to the front of the engine and is driven through a reduction gear drive connected to a second turbine not found in a turbo-jet engine. Unlike the turbo-jet engine, the first turbine drives the compressor, and the second the propeller. About 80 per cent of the useful energy is removed from the exhaust gases in driving the turbines, the rest being used in the form of jet thrust.

Turbo-prop engines are much more efficient below 500 mph than are turbo-jet engines; therefore they find

Diagram of a turbo-prop engine. The main difference between this and the turbo-jet engine is the use of energy from the exhaust stream to drive a propeller, as well as to move the plane by jet action. Courtesy General Motors Corporation.

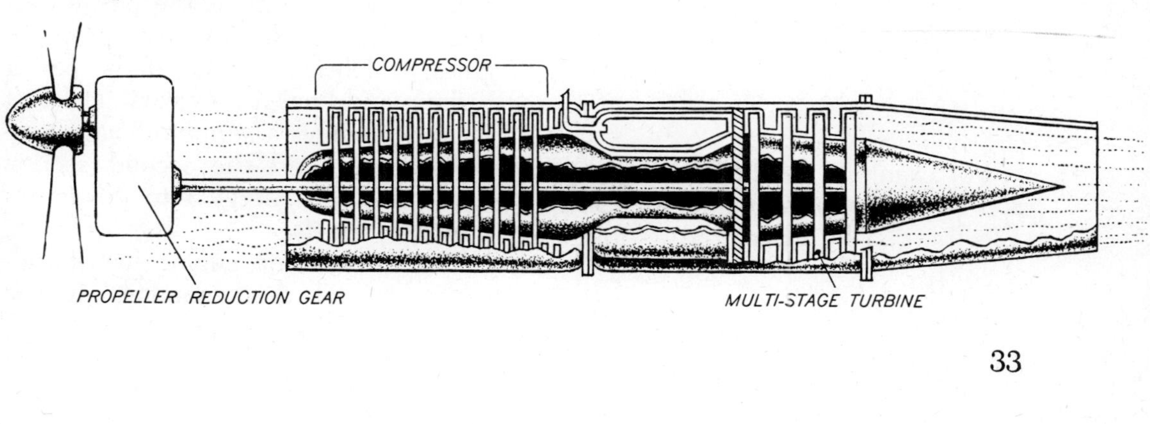

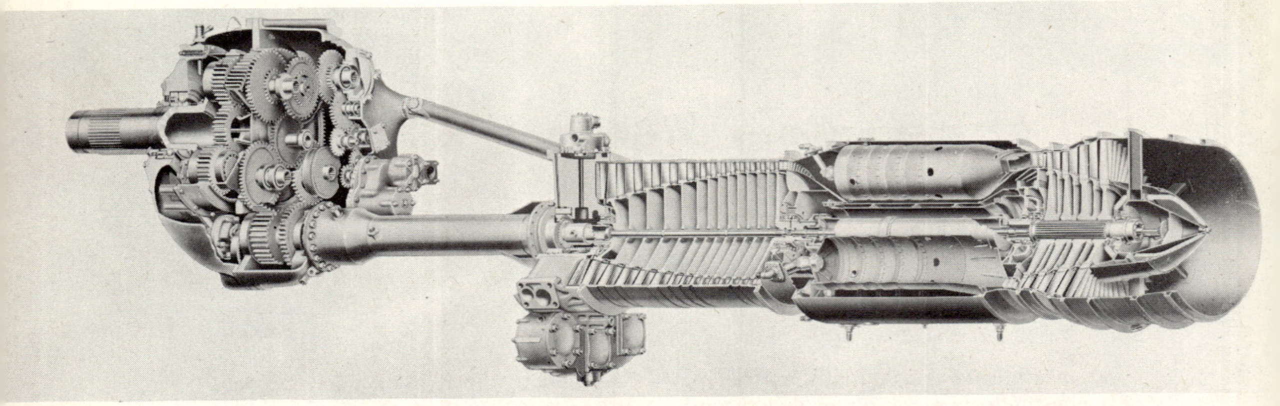

Cutaway view of an Allison 501 turbo-prop engine. See if you can identify the main parts, shown in previous jet illustrations. Courtesy Allison Division, General Motors Corporation.

widespread use for short trips. Turbo-prop engines produce about four times more take-off thrust than turbo-jet engines; therefore, such aircraft can take off on a shorter runway. Because propeller effectiveness is limited by compressibility due to high helical tip speeds, turbo-jet engine propulsive efficiency overtakes turbo-prop engine efficiency above 500 mph.

Turbo-Prop Power Calculations

Power ratings of the turbo-prop engine are stated in terms of Equivalent Shaft Horsepower. This value is determined by converting the exhaust gas thrust into horsepower and adding it to the propeller shaft horsepower to obtain the Equivalent Shaft Horsepower:

$$\text{ESHP} = \frac{T}{2.5} + P$$

T = exhaust gas thrust
P = propeller shaft thrust
2.5 = constant

For example, the ESHP of the Allison 501 turbo-prop engine, used in the Electra, can be calculated as follows:

$$\text{ESHP} = \frac{726}{2.5} + 3{,}460 = 3{,}750 \text{ ESHP}$$

Automobile Turbine Engines

Most automobile manufacturers are experimenting with turbo-jet engines to develop a suitable unit for powering an automobile. Because it is impractical to use the exhaust jet stream directly for motive power, automobile engineers are concentrating on developing turbine sections, coupled to reduction gears, to utilize all the energy possible from the exhaust gases before they are discharged. For this reason, such engines are called turbine engines rather than jet engines. One remaining problem is that of increasing engine operating economy.

An automobile turbine engine contains two sections, the first being the gasifier section, consisting of the usual compressor, combustion chambers, and turbine. Behind this is the power section, similar to aircraft turbo-prop engines, which is made up of the power turbine and final drive gears. The first section provides a supply of hot, compressed gas, and the second section transforms the energy into power for the rear wheels.

The gas turbine engine has its advantages and disadvantages. It can

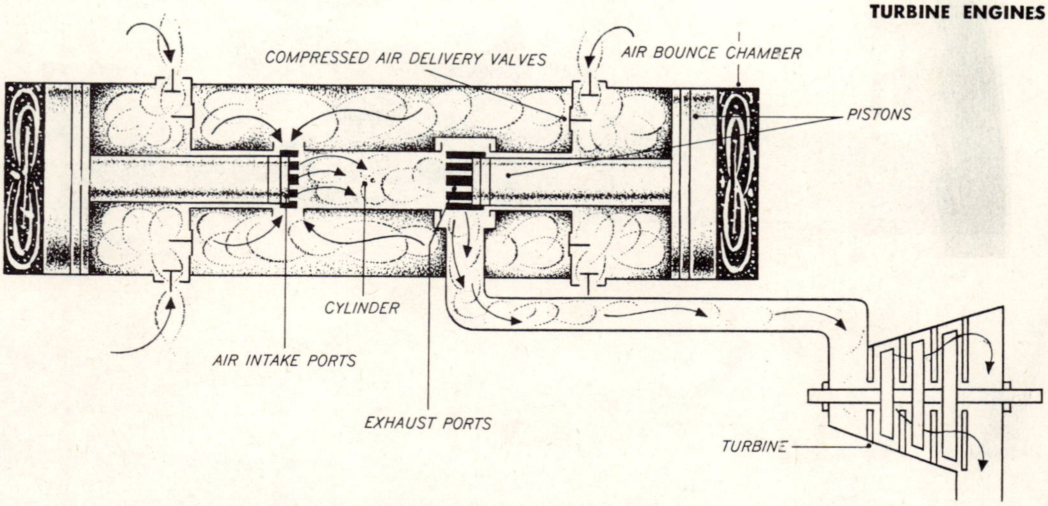

In a free piston engine, the pistons move back and forth by compressing air. Combustion occurs in the center cylinder. The exhaust gases are used to drive a turbine. Courtesy General Motors Corporation.

AS TURBINE

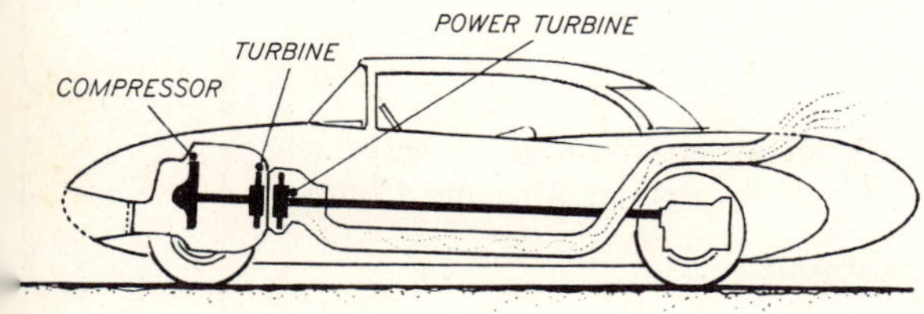

Straight jet engines cannot be used to power automobiles because of the exhaust blast. However, much development work is going into using the exhaust gases to drive a power turbine as shown in this diagram. Courtesy General Motors Corporation.

FREE PISTON ENGINE

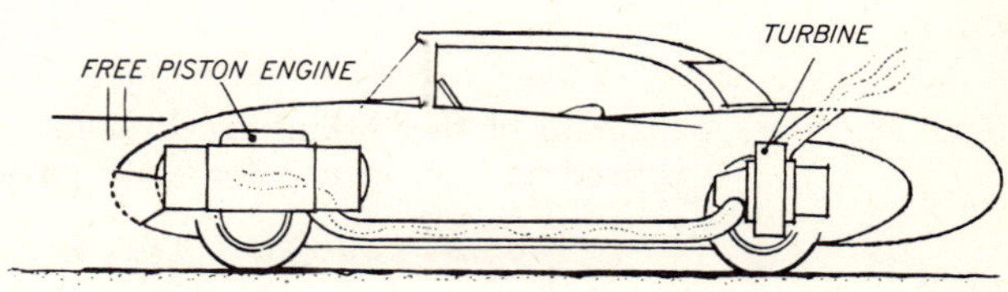

Some manufacturers are experimenting with a free-piston engine. Such an engine compresses air, which is directed to the rear, where a separate turbine drives the wheels. Courtesy General Motors Corporation.

deliver much more power per pound of weight compared with a piston-type engine but it operates efficiently only at top speed; therefore the fuel consumption is much higher when operating in traffic. To increase efficiency, automotive turbine engines use heat exchangers to remove energy from the spent exhaust gases, which is then used to heat the incoming air, thus adding energy to the system.

The "Whirlfire" GT-305 engine, which powers the GM "Firebird" III, is a 225 hp, regenerative gas turbine. It produces power by directing hot compressed gases through a turbine connected to the wheels. Air enters inlet (1, page 38) and is compressed by the turbine (2) to three times atmospheric pressure. It passes through the rotating regenerators (3), absorbing heat, and then enters the combustors

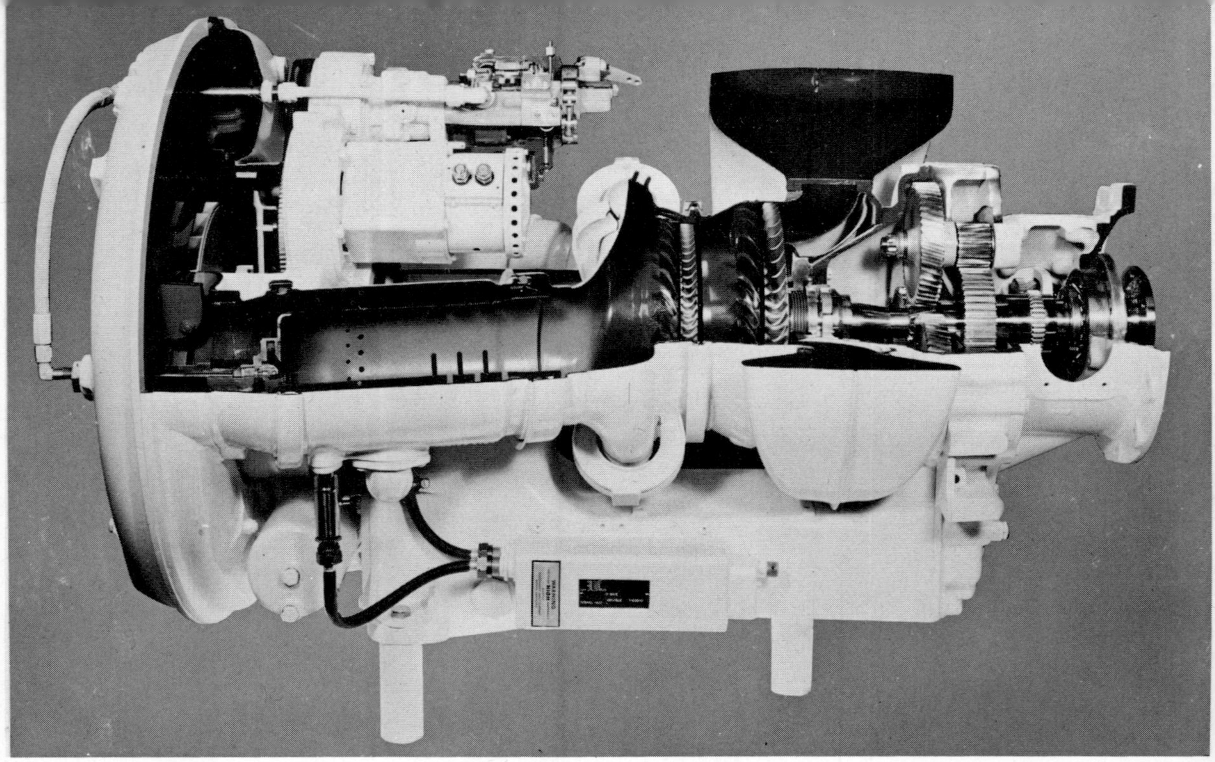

A cutaway view of a turbine engine used in some trucks. Note the two turbines, the right one driving a set of reduction gears. Courtesy Boeing Airplane Company.

A diagram of the turbine truck engine which shows the passage of gases when it is used as a jet engine for small planes. Courtesy Boeing Airplane Company.

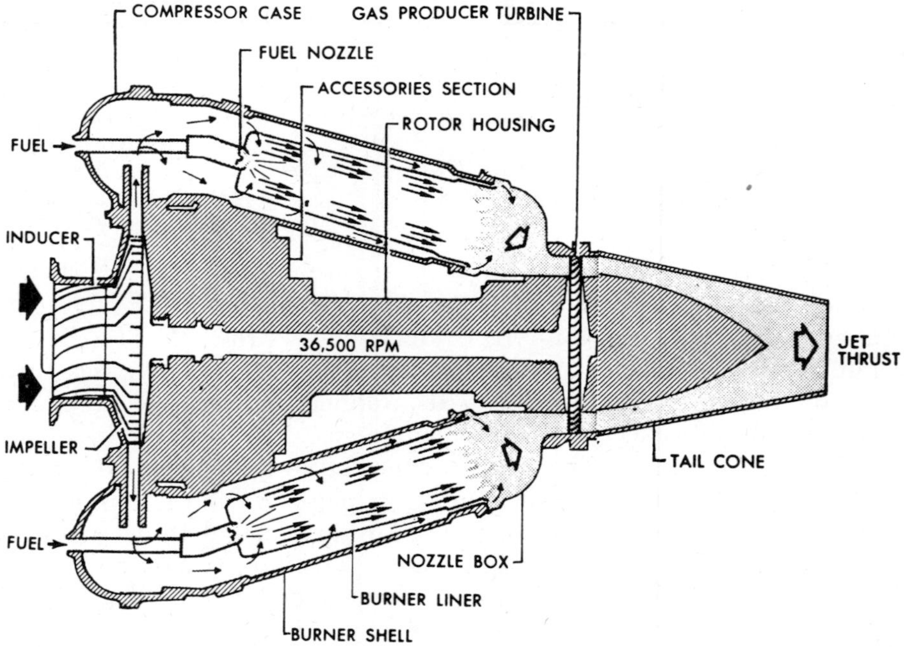

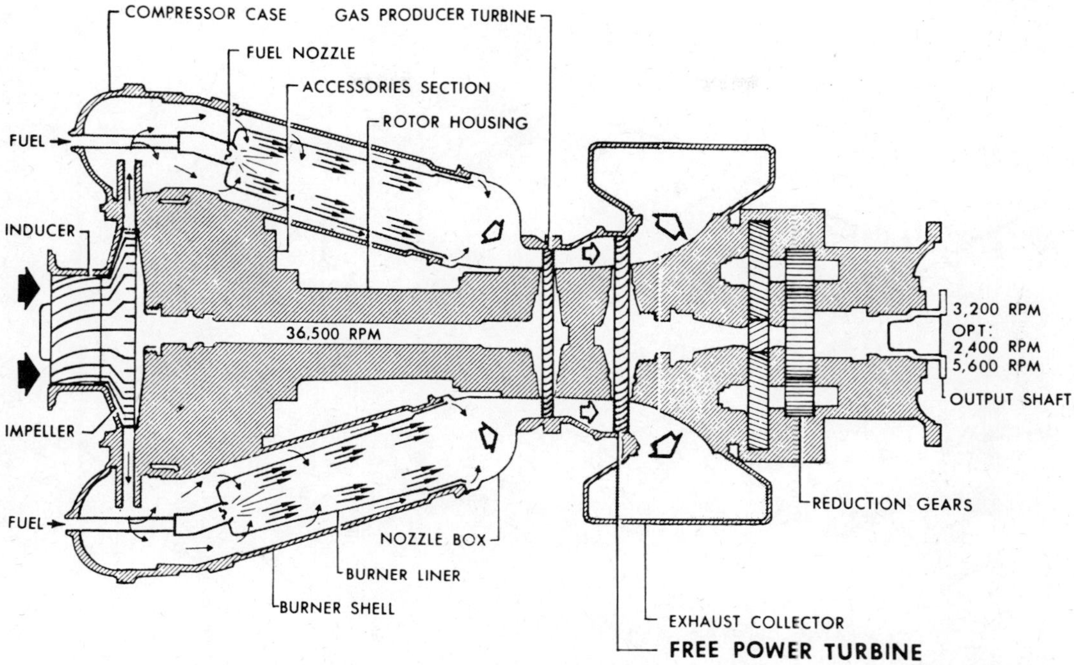

The same engine fitted with a power turbine, for trucks. Courtesy Boeing Airplane Company.

Diagram of the gas turbine engine used to power some advance car models. Courtesy General Motors Corporation.

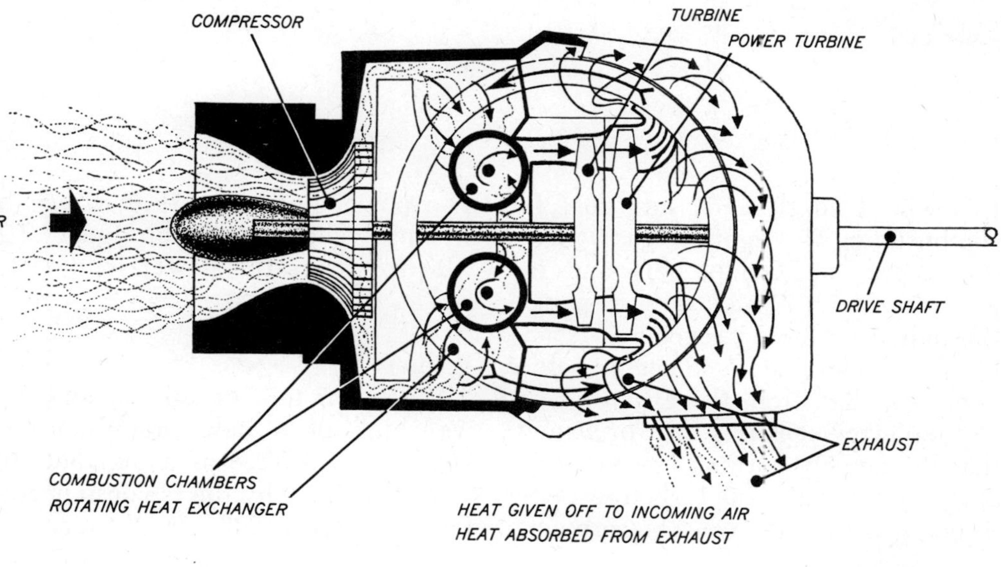

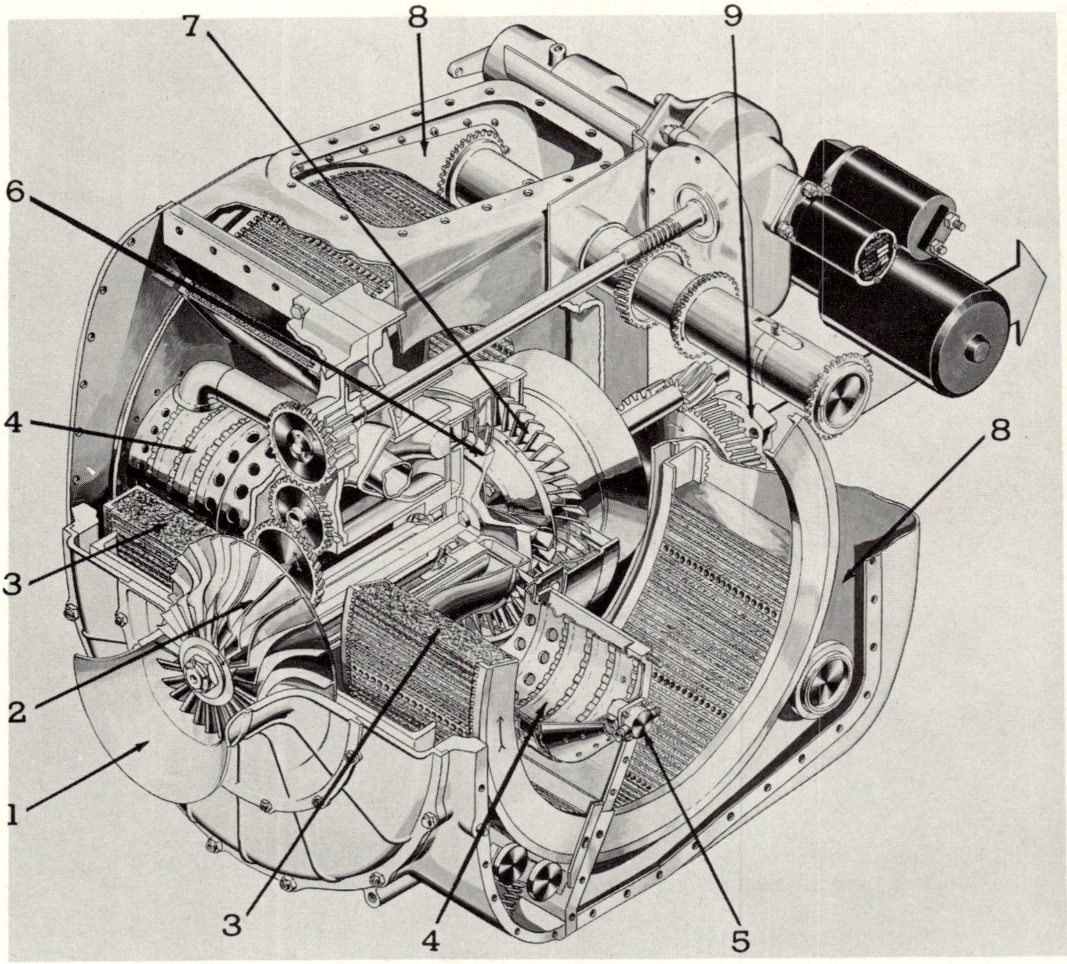

A "Whirlfire" GT-305 gas turbine engine. The numbers refer to numbers in the text. Courtesy General Motors Corporation.

(4) where nozzles (5) add fuel for combustion. The ignited gases drive the compressor turbine (6) and the power turbine (7) and are then passed through the rotating regenerators (3) to absorb its energy (heat) before being expelled from the exhaust ports (8). So efficient is this heat exchanger that the fiercely burning exhaust gases are cooled to 300-500°F. before being discharged from the exhaust ports (8). The power output shaft is driven by a single-stage reduction gear (9) which delivers the power to the rear end.

Rockets

When we talk of rockets and jets, we think of a new propulsion era. However, the records show that rockets were used by the Chinese in warfare as early as 1232 AD. Around 1800,

For many years, rockets have been used by the Army. Courtesy U. S. Army.

A salvo of rockets leaves a navy LSM. Courtesy U. S. Navy.

rockets were so improved by William Congreve (England) that they became an essential part of military equipment. Evidence of American interest in military rockets is provided by the words of a familiar song, written in 1814: ". . . And the rocket's red glare, the bombs bursting in air, Gave proof through the night that our flag was still there . . ."

Rockets were largely replaced during the latter part of the 19th century by large guns which had greater range and better accuracy. Many scientists, however, continued to be interested in man's dream of leaving this planet and experimented with rockets as the motive power. Some of them did leave this planet suddenly in a premature explosion of the engine.

Rockets operate on the same principle as jet engines — Newton's third law of motion: "For every action, there is an equal and opposite reaction." The motive power of a rocket is dependent on the speed of the gas as it leaves the rear of the rocket. This is called the exhaust velocity and is related to the amount of heat produced by the burning fuel.

Rockets differ from turbo-jet engines in that they carry their own oxygen so they are not dependent on the oxygen of the atmosphere. Therefore, they can operate in space.

Rocket Fuels

Basically, a rocket engine depends on oxidizing hydrocarbons to produce a hot flame, just as does any other internal combustion engine. However, because the rocket must carry its own oxidizer (as well as hydrocarbon) much research has gone into finding a fuel which would give the greatest thrust with the least weight.

Initially, black powder was used because it could be compressed into a relatively small space. However, powder develops minute cracks, which increase the burning area so that sometimes the entire mass burns at once and explodes. Some of the latest solid-fuel propellants contain mixtures of nitric acid (HNO_3) and dimethylhydrazine.

Generally, liquid oxygen and alcohol are used as fuels, but they add tremendously to the weight and pumping complexity problems.

Problems of Rocketry

Early problems revolved about the combustion chamber. The temperatures were so high that ordinary steels melted. Problems of accurate guidance bothered rocket builders for many years. The slightest unevenness in burning deflected the exhaust stream and sent the rocket wildly off course. Accurate stabilizers have had to be developed.

Another problem is that of transferring huge quantities of fuel to the rocket motor under a constant pressure and volume, which is necessary to maintain even burning and thrust. The pumps have to be able to empty the tanks (sometimes as much as 50 tons of fuel) within sixty seconds, while handling liquid oxygen at a temperature of 300°F. *below zero.* An auxiliary steam engine is required to operate the pumps, its power generated by concentrated hydrogen peroxide (H_2O_2) and potassium permanganate ($KMnO_4$) which, when brought together, react violently creating steam.

A Jupiter "A" rocket leaves the pad at Cape Canaveral, Fla. Courtesy U. S. Air Force.

A Thor IRBM starts on a journey into space. Courtesy U. S. Air Force.

$$KMnO_4 + 2\,H_2O_2 \longrightarrow 2\,H_2O\uparrow + 3\,O_2 + MnO_2 + K$$

ROCKET GUIDANCE

Early rockets used gyroscopically controlled internal vanes which were forced into the blast of the exhaust to deflect it in order to direct the rocket in the desired direction. Later rockets pivot the motor to direct the exhaust blast at an angle, and some rockets use a combination of these.

Naturally, any vane pushed into the exhaust blast must be made of a heat-resisting material, and graphite was

Launching of a Jupiter ABMA missile. Note the coating of frost from the liquid oxygen which is stored in the missile at 300° F. below zero. Courtesy U. S. Air Force.

found to be one of the materials that withstood such conditions. However, its strength is not great which creates problems of reliability. Much research is being done in this area.

The critical period of rocket guidance is in the first few seconds when the speed is low. Then the internal vanes have to do the entire guidance job. When the speed increases, the external vanes become more effective, but then with the increased altitude the air thins so that the external vanes become less and less effective. Certainly these are not simple problems.

The small jets of flame from the rear end of this Atlas missile are part of the guidance systems being developed. Courtesy U. S. Air Force.

This Honest John missile shows three jets of flame from its sides which are part of the guidance system. Courtesy U. S. Army.

This Matador is very similar in shape to the German V-1 rocket used in World War II. Courtesy U. S. Air Force.

Firing Sequence

The firing of one of the early German rockets clearly shows its operation in three distinct stages as follows: First the fuse was ignited. A shower of sparks from the exhaust tube indicated that the fuse was burning. Then the fuel tank valves were opened so fuel could enter the motor by gravity. The initial thrust was about 7 tons, not enough to move the 14 ton rocket from the pad. However, this initial firing was a test of motor operation, after which the auxiliary pumping engine was energized by auxiliary fuels and began to pressure-feed the rocket motor. In about three seconds, the rocket thrust jumped to about 27 tons which forced the rocket slowly into the air. During the first second, the rocket moved about its own length. Since the rocket consumed 280 pounds of fuel every second, the acceleration built up 35 feet per second. As the rocket reached higher altitudes, the lower air pressure allowed the exhaust velocity to increase which also increased the thrust. At a height of ten miles, this factor alone added 4.5 tons of thrust.

War Influence

Rocketry took a huge step forward in World War II. Germany developed the V-1, not a true rocket because it contained a pulse-jet engine powered by gasoline, taking its oxygen from the air. The Germans used a steam-propelled catapult to start the V-1 on its way. The forward motion of the missile opened a set of shutters on the front so air could be compressed into the combustion chamber. At the same time, gasoline was sprayed into the chamber and ignited. The resulting expansion closed the shutters, forcing the gases from the rear of the engine. As soon as combustion pressure died down, the forward motion of the missile opened the shutters to admit more air and repeat the cycle. The V-1 flew at about 360 mph.

Later, the Germans developed the V-2 which was a true rocket, 47 feet tall and weighing a little over 12 tons at take-off. The warhead weighed one ton, the body of the rocket three tons, and the fuels which were liquid oxygen and alcohol, weighed about 8 tons.

First Step Into Space

It soon became apparent that a single rocket would not be able to hold enough fuel to propel itself out of the earth's gravitational field. Therefore, two rockets were put together so that the first one was a booster for the second. When the first rocket consumed all its fuel, it dropped away from the forward rocket, which ignited

This dramatic picture shows the launching of a two-stage missile, a Thor-Able. Courtesy U. S. Air Force.

after having gained the velocity and altitude of the first rocket.

By 1954, scientists began to plan for an orbit shot. They reasoned that if they could send a rocket fast enough, the earth's gravity would not be able to overcome its forward motion to drag it back again. If the speed and angle were correctly calculated, gravity would bend the rocket's path so that it would travel around the earth, balanced in orbit by centrifugal force. The orbital velocity is 18,000 mph at 300 miles altitude and scientists have calculated that it takes a three-stage rocket to provide the required thrust.

On October 4, 1957, the Russians put the first artificial satellite into an elliptical orbit which at its perigee (closest point) was 156 miles from the earth and at its apogee (farthest point), 560 miles. Because part of the orbit was inside of the earth's atmosphere, the satellite gradually slowed from friction with the air and came closer to the earth at each pass. Eventually, it fell to earth and burned in January, 1958.

Some interesting facts about the American rocket, Explorer, which was sent into orbit on January 31, 1958, are that the rocket consisted of four stages. The first stage burned for about two minutes, providing a velocity of 6,800 mph. At 53 miles altitude, the burned-out first stage dropped off, and the other three stages were ignited one after the other to place the satellite into an orbit which was 220 miles at its perigee and 1,600 miles at its apogee.

The next step forward will be to build a space station so that man can take off to all points of the Universe without first having to combat gravity and the friction of the atmosphere. Truly, the next few years will be busy ones and interesting to observe.

CHECK YOUR KNOWLEDGE

1. How is the sun's energy generated?
2. What is the velocity of light?
3. Explain the statement, "If man could learn to get power directly from sunlight, the world would be changed overnight."
4. What is photosynthesis?
5. What is considered man's first great step forward in the development of power?
6. What is an aeolipile? How does it operate?
7. What is nuclear energy?
8. Describe Newcomen's steam engine.
9. Describe the operation of a piston-type steam engine.
10. What is an external combustion engine?

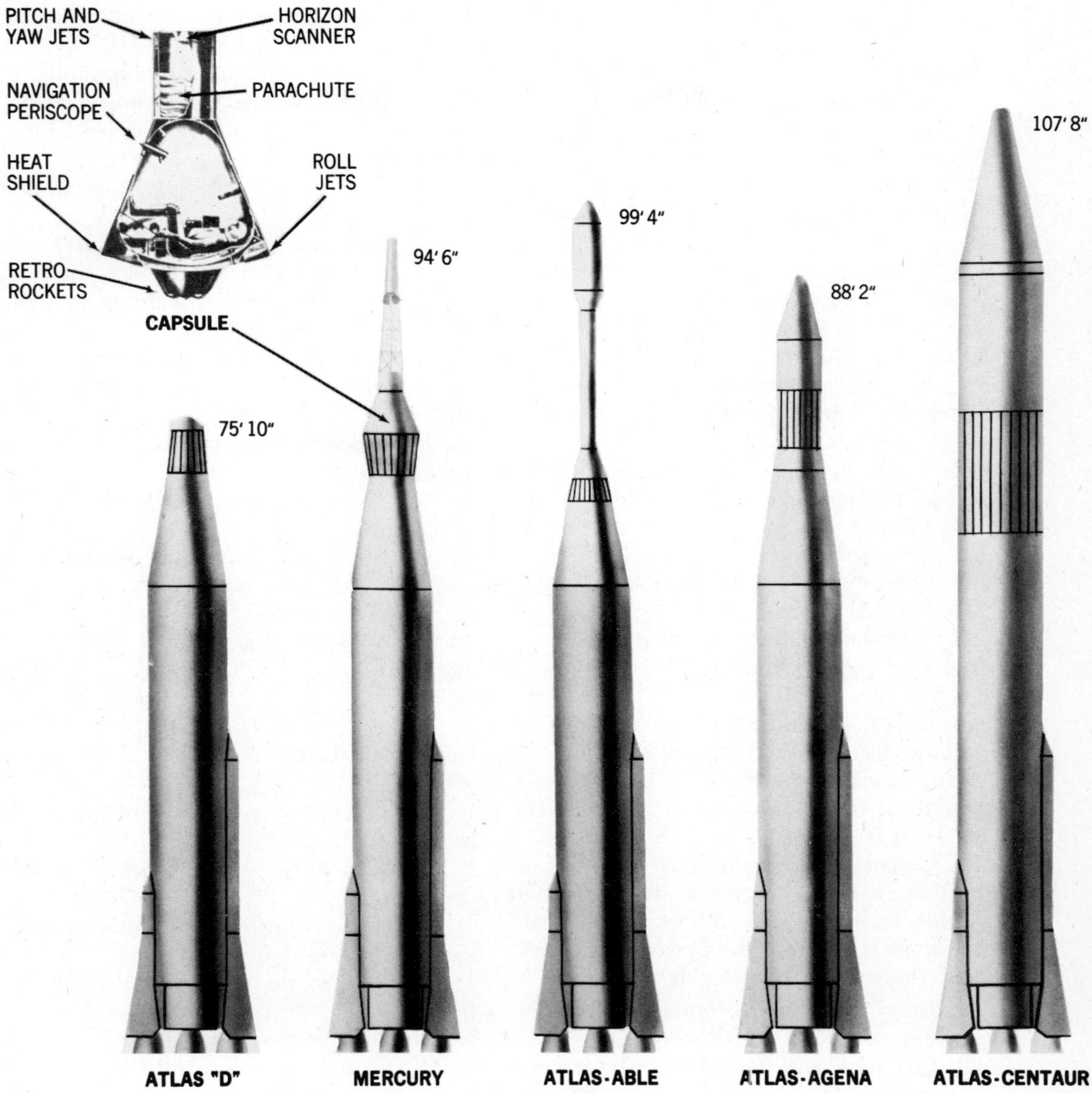

This shows to scale the Atlas family of space vehicles. The Mercury vehicle is the one that carried Commander Alan B. Shepard, Jr., U. S. N., the first American spaceman, into outer space. Courtesy U. S. Air Force.

11. Describe the operation of a steam-turbine engine.
12. In what way is a gasoline engine like a cannon?
13. Describe the operation of Huyghens' engine.
14. Describe Otto's improvements to Huyghens' engine.

15. What is the Otto cycle?
16. What is the purpose of a crankshaft and connecting rods?
17. How long does it take the flame front to pass from one end of the combustion chamber to the other in a gasoline engine?
18. How does the two-stroke cycle gasoline engine differ from the four-stroke cycle engine?
19. Describe the operation of the two-stroke cycle gasoline engine.
20. On what principle is ignition based in a diesel engine?
21. List the differences between the strokes of a four-stroke cycle gasoline and diesel engine.
22. Who developed the two-stroke cycle diesel engine?
23. What is the purpose of the air pump in a two-stroke cycle diesel engine?
24. Describe the operation of a two-stroke cycle diesel engine.
25. Calculate the S.A.E. horsepower rating of an eight-cylinder engine with a four inch bore.
26. Calculate the brake horsepower of an engine, using a Prony brake in which the force measured on the scale is 50 lbs., the length of the lever arm two feet, and the speed 1,000 rpm.
27. Calculate the brake horsepower of an engine, using a Prony brake with two scales in which the difference in scale readings is 100 lbs., the flywheel two feet, and the engine running at 1,000 rpm.
28. How many watts equal one horsepower?
29. List five advantages of turbo-jet engines over piston-type engines for aircraft use.
30. How does a jet engine operate?
31. What force pushes a jet engine forward?
32. How are the flames of combustion kept from striking the walls of a jet engine combustion liner?
33. What is an afterburner? How does it operate?
34. How does a jet engine ignition system differ from a piston-type engine ignition system?
35. How is the afterburner fuel ignited?
36. In what way does the turbo-prop engine and turbo-jet engine differ besides the obvious propeller?
37. What advantage has a turbo-prop engine over a turbo-jet engine?
38. In what way does an automotive jet engine vary from an aircraft jet engine?
39. What is the great disadvantage of a jet engine for automotive use?
40. How does a heat exchanger operate?
41. On what principle does a rocket operate?
42. How does a rocket differ from a jet engine?
43. What are the commonly used rocket liquid fuels?
44. On what chemicals does the auxiliary pumping engine operate?
45. How is a rocket guided?
46. Describe the operation of a pulse-jet engine.

Self-Check Tests

TRUE-FALSE

1. Nuclear energy is the basis of the sun's heat.
2. The velocity of light is 186,000 miles per minute.
3. The aeolipile utilizes the same operating principle as does a jet engine.
4. The first steam engine was built by James Watt.
5. A turbine engine has pistons.
6. All steam engines have pistons.
7. Christian Huyghens' engine operated on vacuum rather than pressure.
8. In a four-stroke cycle gasoline engine, the piston moves up the cylinder during the intake stroke.
9. A two-stroke cycle engine fires everytime the piston reaches the top of its stroke.
10. The two-stroke cycle gasoline engine is the only one using a reed valve to control the flow of carbureted mixture.

11. The gasoline engine has a higher compression ratio than does a diesel engine.
12. In both the gasoline and diesel engines, the output is controlled by the amount of fuel fed to the cylinders.
13. Horsepower is a rating given the ability of an engine to do work.
14. A turbo-jet engine develops its thrust by pushing against the air behind it.
15. Because an afterburner is located behind the turbine of a jet engine, it does not increase fuel consumption appreciably.
16. An afterburner is never used with a turbo-prop engine.
17. An afterburner requires a separate spark plug for ignition.
18. The horsepower output of a turbo-prop engine can be calculated by dividing the thrust by the constant 2.5.
19. The great advantage of the jet engine for automotive use is that it has less weight per horsepower; therefore, it operates more economically at low speeds compared with a piston-type engine.
20. Black powder is the best solid-fuel propellent for rockets.
21. Some rockets are steered by tilting the motor.

MULTIPLE CHOICE

1. Of the sun's energy which reaches the earth, about: a) ten per cent; b) one per cent; c) 1/10th of one per cent; d) 1/100th of one per cent is stored in the form of fuel.
2. The process by which the sun changes carbon dioxide and water into carbohydrate and oxygen is called: a) combustion; b) oxidation; c) photography; d) photosynthesis.
3. The first steam engine was built by: a) Thomas Newcomen; b) James Watt; c) Christian Huyghens; d) Nicholas Otto.
4. The first gasoline engine to operate from expanding gas was built by: a) Thomas Newcomen; b) James Watt; c) Christian Huyghens; d) Nicholas Otto.
5. In a four-cycle gasoline engine, the exhaust valve starts to open during the: a) compression; b) power; c) exhaust; d) intake stroke.
6. A two-stroke cycle gasoline engine contains: a) poppet valves; b) blower; c) reed valve; d) injector.
7. A four-stroke cycle diesel engine has a: a) blower; b) reed valve; c) carburetor; d) injector.
8. A two-stroke cycle diesel engine contains a: a) blower; b) reed valve; c) carburetor; d) spark plug.
9. One horsepower is equivalent to consuming: a) 467; b) 647; c) 476; d) 746 watts of electric power.
10. A jet engine combustion chamber liner is kept cool by an: a) air stream; b) liquid coolant; c) heat exchanger; d) radiator.
11. A jet engine uses: a) gasoline; b) liquid oxygen; c) kerosene; d) high octane fuel.
12. The jet engine ignition system contains a: a) magneto; b) distributor; c) condenser; d) ignition by compression.
13. Turbo-prop engines are more efficient than turbo-jet engines for aircraft flying at: a) high altitude; b) over 500 mph; c) long trips; d) short trips.
14. Which of the following is used *only* on automobile, turbine-type engines: a) compressor; b) turbine; c) heat exchanger; d) afterburner?
15. Rockets were invented by the: a) Chinese; b) Egyptians; c) Germans; d) Americans.

COMPLETION

1. When we burn a piece of wood, we ________ carbon and hydrogen.
2. Today, most of our power is generated ________ from the sun.
3. All engines are of ________ basic types.
4. The turbine engine is sometimes referred to as a ________-type engine because it derives its power indirectly.
5. In a ________ engine, ignition is by compression.
6. The ________ ________ ________ engine

fires every time the piston reaches the top of its stroke.
7. Jet engines operate on Newton's .... law of motion.
8. An .... increases turbo-jet engine power output about 1/3.
9. Ice in the .... restricts the flow of fuel so that a "flame-out" results.

MATCHING

1. External combustion
2. Turbine engine
3. Four-stroke cycle engine
4. Two-stroke cycle engine
5. Ignition by compression
6. Afterburner
7. Heat exchanger
8. Rocket
9. Shutters

1. Nicholas August Otto
2. Steam engine
3. Turbo-jet engine
4. Pulse-jet engine
5. Carries own fuel oxidizers
6. Diesel engine
7. Intake and exhaust ports
8. Turbo-prop engine
9. Automobile turbine engine

Ancient man noted that a piece of loadstone always pointed to the North Pole.

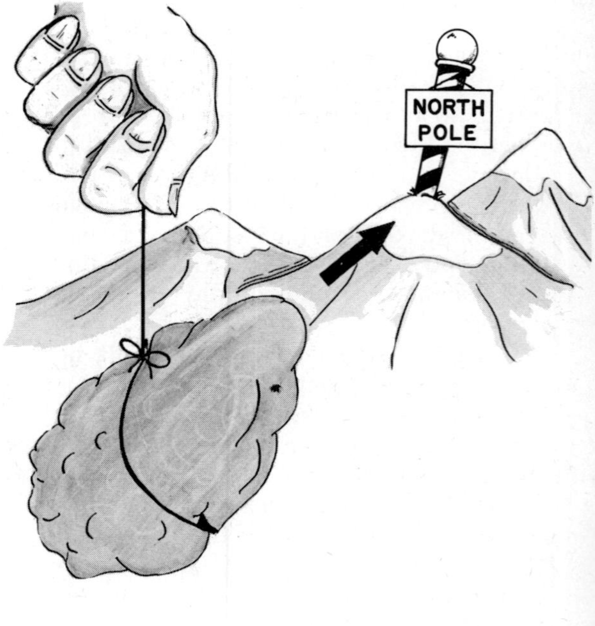

CHAPTER 2

FUNDAMENTALS OF ELECTRICITY

MAGNETISM

Magnetism is the basis of electricity. Experiments have shown that an electric current can produce magnetism, and magnetism can produce electricity. Many electrical effects are explained in magnetic terms. Therefore a knowledge of magnetism is important to understand electricity.

Magnetism was first noted by a man of ancient times, who found that a certain kind of rock always pointed in the same direction when suspended on a string. This natural magnetic substance was called *loadstone.* It was used as the first crude compass. Later investigation proved that the earth itself is a gigantic magnet, and its attraction for the loadstone caused the first inquiries into magnetic properties.

The present theory of magnetism is based on the idea that each molecule of a piece of certain materials is in itself a small magnet. In an unmagnetized piece of iron, the molecules are freely arranged. Magnetization consists of turning all these small magnets in the same direction, so that they combine to produce a powerful force.

Magnets are either of the permanent or temporary types. Once the molecules of hard steel are forced to line up in the same direction, they tend to maintain this position. Such a magnet is called a permanent magnet. When a piece of soft steel is placed in a magnetic field, the molecules line up as in hard steel but return to their original positions as soon as the external force is removed. Hence soft steel is used to make a temporary magnet.

In a magnetized piece of steel, the molecules align themselves in the same direction and combine their forces into a single magnet. The molecules of an unmagnetized piece of steel point in every direction.

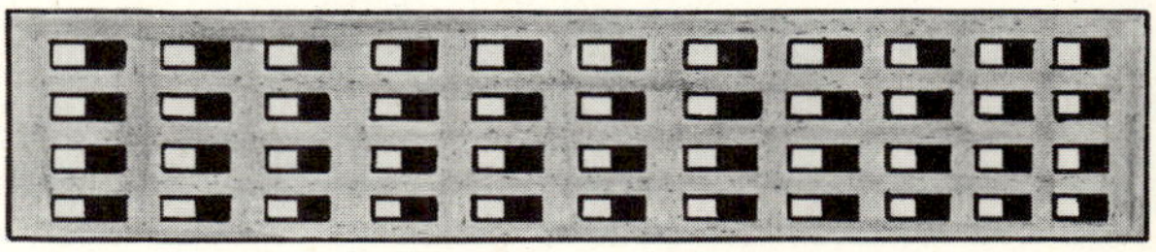

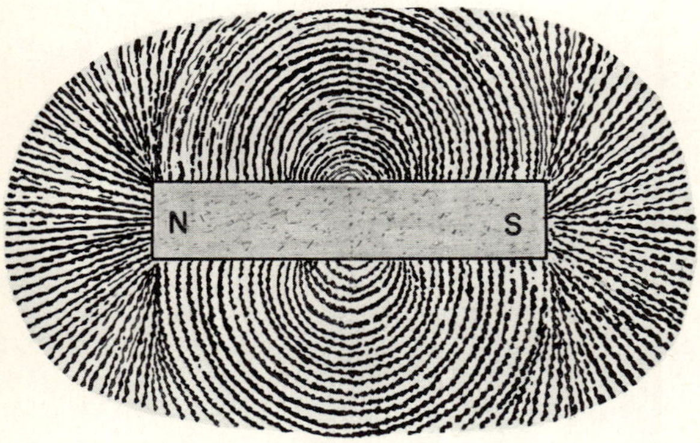

The field of force extends out from each pole of a magnet in circular lines.

MAGNETIC POLES AND FIELDS

When magnetized, a piece of steel has a field of force around the ends, which are called *poles*. Because a freely suspended magnet rotates until it lines up with the earth's geographic poles, the end pointed toward the earth's North Pole is called a north-seeking pole or, more commonly, the *north pole* of the magnet. The opposite end of the magnet is called a *south pole* because it points toward the South Pole of the earth.

Like poles repel and unlike poles attract.

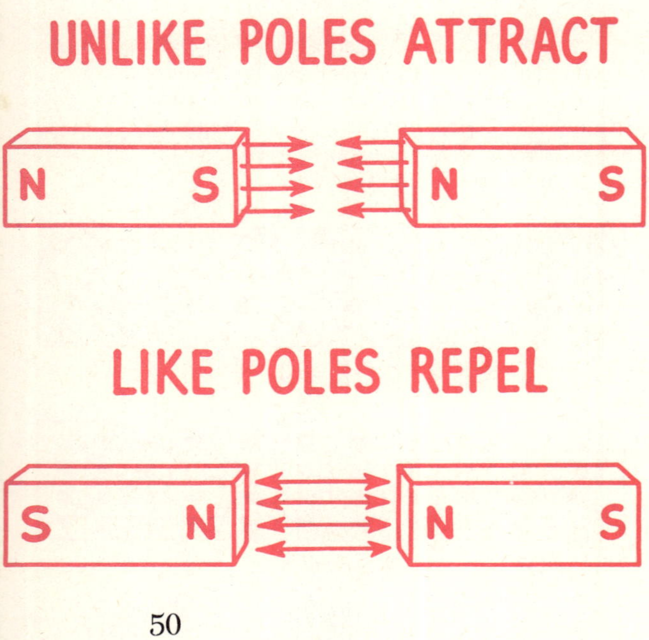

If two freely suspended magnets are brought close together, they will swing until a north pole of one is opposite a south pole of the other. Then the magnets will attract each other. Should you turn one magnet so that similar poles are facing each other, the magnets will be repelled. From these facts the *law of magnetism* is derived: namely, *like poles repel,* and *unlike poles attract.*

The region around the poles of a magnet is called a *field of force.* If a piece of steel is placed within this field, it will be attracted to the pole. If a compass needle is placed near the pole, the needle will swing toward it. While we cannot see a field of force, its effects can be shown by the compass.

BATTERIES

A single-cell battery is a container of acid in which two dissimilar metals are submerged. The chemical action of the acid on the metals sets up a flow of electricity. *Dry cells* contain the acid in the form of a paste or jelly.

When two plates of a single cell battery are connected together, a complete circuit is made, and electrons flow. The two plates of the battery are called poles. The electrons, being negative particles of matter, flow from the negative pole to the positive pole.

Electrons will flow along a circuit only when they are pushed by pressure. It is the action of the acid on the metals within the battery which creates this pressure, called *potential difference* or *volts.* The pressure of a fully charged wet cell battery is 2 volts, while that of a dry cell is 1½ volts.

The physical size of a dry or wet cell

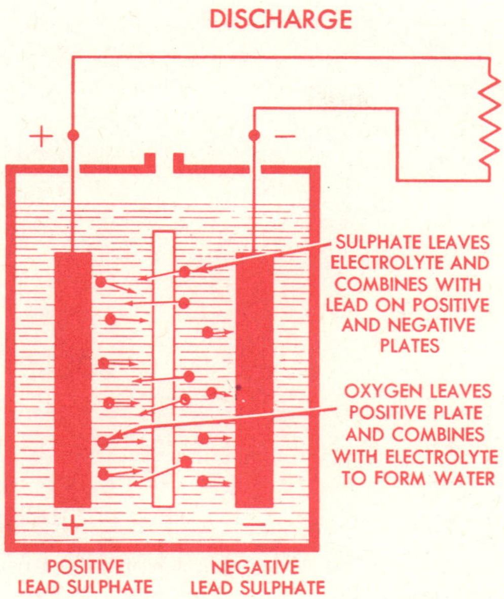

A wet cell battery contains two dissimilar metals submerged in an acid solution. Courtesy Lincoln Division, Ford Motor Company.

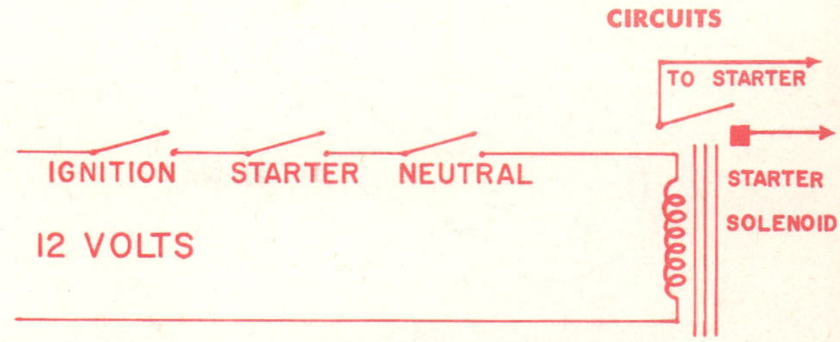

In a series circuit, the current must flow through each switch in turn.

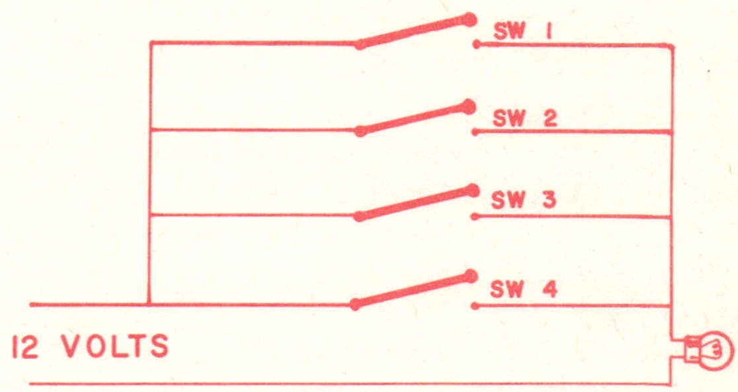

In a parallel circuit, the current can select any one of several paths.

has no effect on the pressure or voltage. However, a large cell contains much more active material and will last longer under similar service. In other words, a large cell, wet or dry, supplies a larger number of electrons before the cell is exhausted.

The quantity of electrons flowing in a circuit is called *amperage*. This is discussed later.

CIRCUITS

Electricity can flow only when there is a complete circuit. That is, there must be an uninterrupted path from the electrical source to the device being operated and back to the source. Otherwise the circuit is said to be *open*.

Two types of wiring circuits are commonly used in the automobile—*series* and *parallel circuits*. In a series circuit, *all* the current leaving the battery must pass through *all* the electrical devices in that circuit. In a parallel circuit, there may be several branches so that only a part of the electricity flows through any one branch.

The starter and ignition switches are usually connected in series, in which case the ignition switch must be turned on before the starting motor switch will work. Some four-door cars are equipped with four door switches in parallel. Opening any one door lights the interior light.

Batteries are frequently connected in series to increase the total voltage. The length of time that two such batteries can supply current is the same as if one battery were used. Batteries are sometimes connected in parallel to increase service life. The voltage of batteries hooked in parallel remains the same as for one battery.

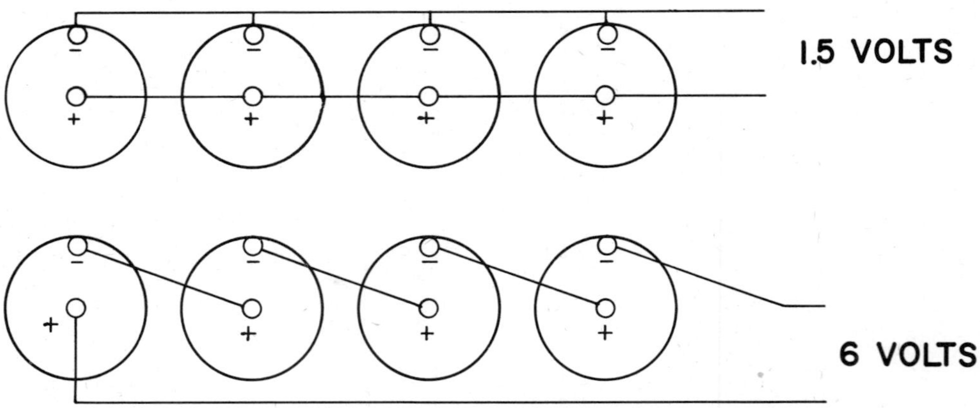

Dry cells are connected in parallel to increase the amperage or capacity. They are hooked in series to increase the voltage.

The automotive storage battery contains both series and parallel connections. Within each cell several plates are connected together in a parallel circuit to furnish enough current for the starting motor. Each of the six cells (2 volts each) is connected in series to furnish twelve volts to energize the other electrical equipment.

ELECTROMAGNETISM

An electromagnet consists of a coil of wire wound around a soft iron core and connected to a source of current. Electromagnets depend on the flow of electrons in the coil of wire for their force and can be "turned off" by shutting off the electricity. Electromagnetism is used in automobiles, boats, aircraft, space missiles, and home equipment to operate electric motors, relays, and solenoids. An electromagnet can exist without the iron core but it will be weaker. The core concentrates the lines of force in the center of the magnet. The strength of an electromagnet depends on the amount of current flowing in the coil and the size and kind of core material; the strength is classified according to the number of *ampere turns* it contains. For example, an electromagnet of a 100 turns of wire, allowing one ampere to flow, is equal in strength to another coil of 50 turns allowing two amperes to pass. Both contain *100 ampere turns.*

An electromagnet is a coil of wire with a soft iron core which, when energized with electricity, exhibits the same properties as a permanent magnet.

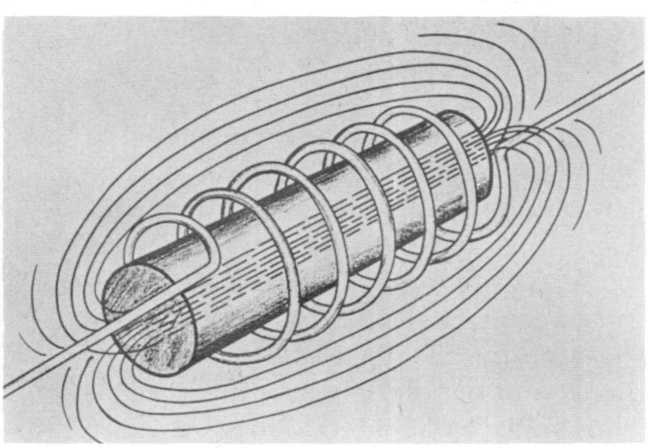

The polarity of an electromagnet can be predicted by means of the *left-hand rule.* If a coil of wire is grasped with the left hand, so that the fingers point in the direction of electron flow, the

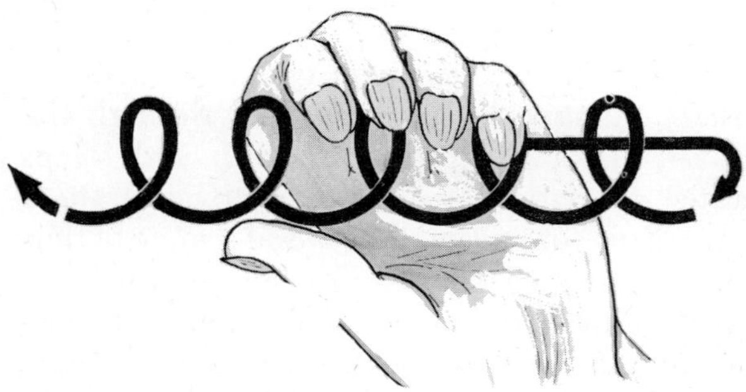

By holding a simple coil with your left hand, your thumb pointing in the direction of the north pole, your fingers will show the direction in which electrons flow.

thumb will point to the magnet's north pole. Or, if the north pole is known, but you wish to determine the direction of electron flow, place your left hand on the coil so that your thumb points in that direction; then the electrons flow in the direction in which your fingers point.

The iron core will move to the center of a solenoid when the coil is energized.

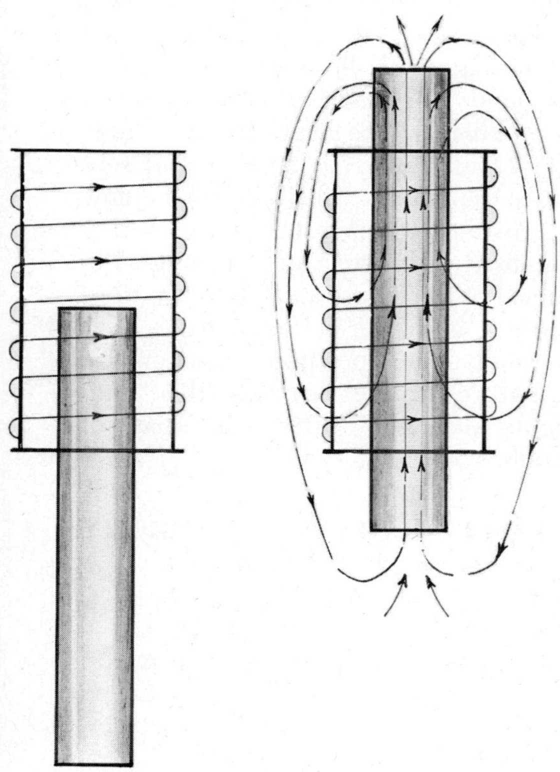

SOLENOIDS AND RELAYS

A *solenoid* is an electromagnet with a movable soft iron core. When the coil is energized, the iron core moves to the center. This movement is used to operate switches and levers.

In the starting motor, a solenoid moves the starter gear into mesh with the flywheel, and then closes the switch which sends electricity to the starting motor. In an overdrive transmission, a solenoid is used to control the shift mechanism.

A *relay* is an electromagnetic switch. It contains a coil of wire and a soft iron core to form the magnet; above this is a hinged piece of flat metal called the armature, which is attracted toward the core when energized. The end opposite the hinge contains a set of contact points used to control another electrical circuit.

Turning the ignition switch in the driver's compartment past the IGN position directs electricity to the relay coil, making it a magnet. It, in turn, energizes the solenoid coil, which shifts the starter gear into mesh with the flywheel ring gear and completes the starting motor circuit to crank the engine. (Some manufacturers do not employ the relay.)

You might wonder why it takes three switches to operate the starting motor – the ignition switch in the driver's compartment, the relay, and the solenoid. The high amperage required to crank the engine cannot be carried through the small switch that the driver activates, so it is used only to operate the relay. The relay controls the solenoid because its heavier contacts can safely carry the larger amounts of current needed to energize the solenoid coil. The solenoid, by its mechanical movement of the plunger, completes the starting motor circuit through very large contacts.

CIRCUIT BREAKER

A circuit breaker is a combination relay and solenoid. Used in place of a fuse, it has the advantage of not having to be replaced when a short circuit occurs.

Some circuit breakers contain a movable iron core which centers itself within the solenoid coil when an excessive amount of current passes through the unit. The core moves up and forces the armature above it to separate a set of normally closed points, thus breaking the overloaded circuit. This eliminates the trouble temporarily because no current flows in the damaged circuit, and so the circuit breaker resumes its normal position, allowing the contacts to close. If the short circuit remains, the cycle repeats itself continuously.

Wiring diagram of a circuit breaker. Note that the contact points are in series with the relay winding. Courtesy Delco Remy Division, General Motors Corporation.

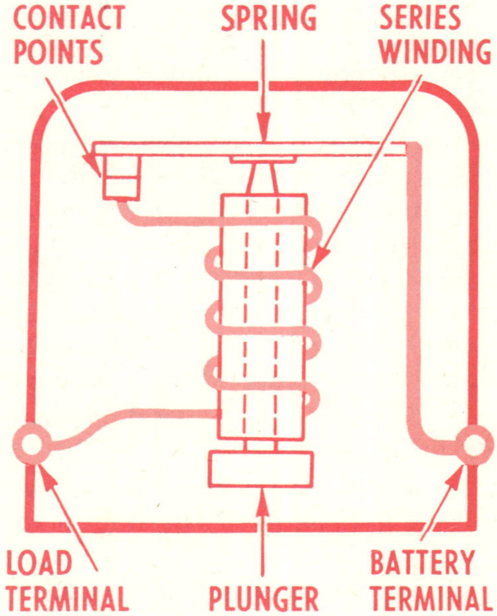

Some circuit breakers contain a *resistance wire* connected across the contact points so that, when the points are open, some current flows through the resistance wire, causing it to heat. The heated resistance wire is wound around a bimetallic armature which bends and keeps the points separated as long as the wiring defect exists.

Some circuit breakers are designed around a bimetallic switch instead of a solenoid. An abnormal flow of current through this type heats the bimetallic blade sufficiently to bend it and separate the points. This reduces the flow of current, which protects the wiring against overheating and burning. Removal of the excess load cools the thermostat, and the points resume their normally closed position. In action, the armature vibrates, causing the headlights to flicker, warning the driver of trouble.

MEASUREMENTS OF ELECTRICITY

Voltage is electrical pressure. It is what "pushes" electrons along a wire. Electrons flow only when there is a difference in pressure. Just as water flows from a point of high pressure to one of low pressure, so do electrons. The unit for measuring electrical pres-

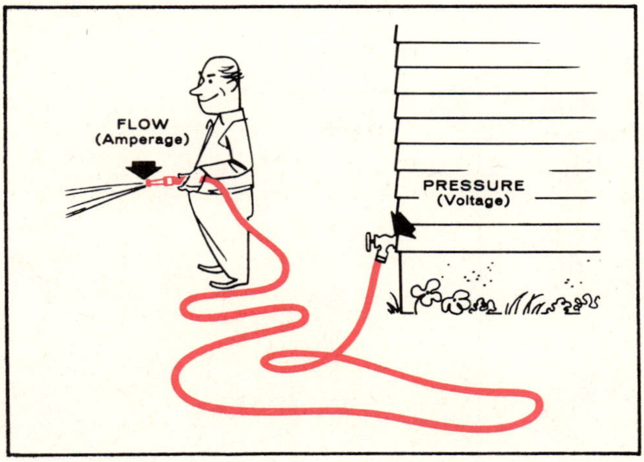

Voltage is pressure. Courtesy Delco Remy Division, General Motors Corporation.

sure is the *volt* and the unit for measuring the quantity of electrons flowing is the *ampere*.

Resistance is electrical friction. It hinders the free movement of electrons in a circuit. The more resistance, the fewer electrons flow. The unit for measuring electrical resistance is the *ohm*. One ohm is the amount of resistance in a circuit which allows one ampere of current to flow under a pressure of one volt.

Resistance depends on the length, size, kind of material, and temperature of a wire. The longer the wire, the more resistance it has; also, the thinner the wire, the more resistance and, when the temperature is increased, the resistance rises rapidly.

Some materials are far better conductors of electricity than are others. Metals like copper and silver are good conductors; therefore their resistance is low. Materials which do not conduct electricity at all are called insulators. Naturally, their resistance is extremely high.

OHM'S LAW

The flow of electrical current is retarded by resistance and increased by voltage. This simple relationship is the basis for Ohm's law discovered by G. S. Ohm (Germany) in the early 19th century. One volt forces one ampere through a circuit with one ohm resistance, and two volts force two amperes through the same circuit of one ohm resistance. This can be expressed by the formulas:

Amperes = volts divided by the resistance
Resistance = volts divided by the amperes, OR
Volts = resistance times the amperes

Resistance retards the flow of amperage. Courtesy Delco Remy Division, General Motors Corporation.

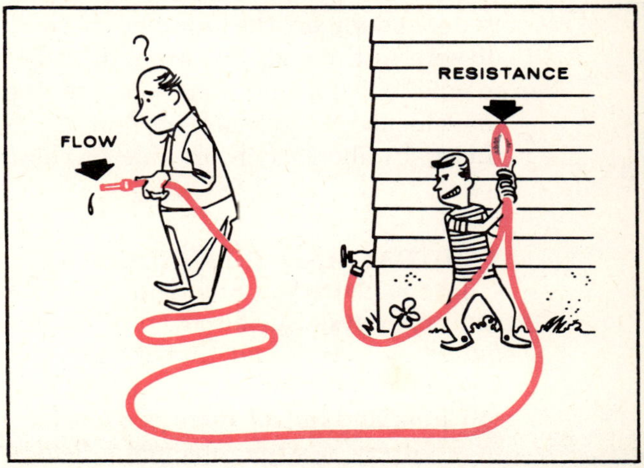

For example, suppose you want to calculate the resistance of a circuit in which a 1.5 volt dry cell battery is connected to light a small bulb which consumes .25 ampere of current. The formula to use is Resistance=Volts divided by Amperes, which, in this case, is:

$$\text{Resistance} = \frac{1.5}{.25}, \text{or 6 ohms.}$$

INDUCED ELECTROMOTIVE FORCES

Michael Faraday (England), a chemist and physicist of the 18th-19th centuries, discovered that if he moved a coil of wire in a strong magnetic field, electric current flowed in the wire. He proved this by connecting both ends of the coil to a sensitive instrument which we call a *galvanometer.*

The current is produced by movement. It makes no difference whether the coil or the magnetic field is moved —but movement must occur. The process by which electric currents are generated in a wire, through its movement in a magnetic field, is called *induction.*

The amount of induced electricity is dependent on the number of turns of wire in the coil, the strength of the magnetic field, and the speed of the movement. A stronger current is induced by increasing any one of these items.

MUTUAL AND SELF-INDUCTION

Joseph Henry (America), a 19th century physicist, discovered *mutual induction.* He found that if two coils of wire are wound around a piece of soft iron, and one of them is connected to a battery, a strong magnetic field is

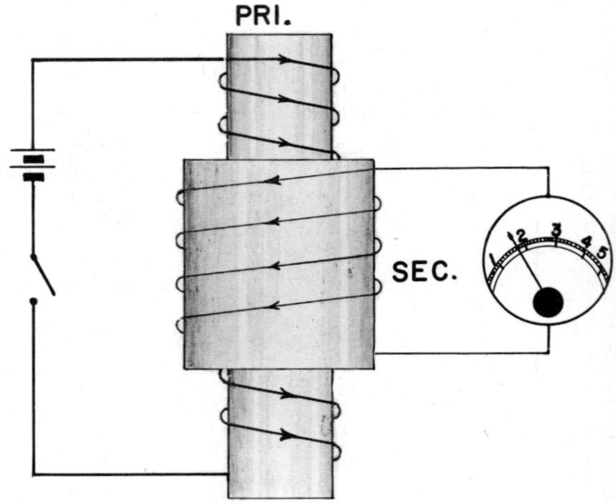

When the primary coil is energized, voltage appears across the secondary coil.

produced. The lines of force created by the magnetism of the first coil (called the *primary*) move rapidly across the wires of the second coil (called the *secondary*), causing an induced current to flow. This process is called *mutual induction.* However, the secondary current makes the secondary coil a magnet, causing it to induce a second flow of current in the primary coil which is in a direction opposite to the initial flow. The generation of this reverse current is called *self-induction.*

A *transformer* is made up of two such coils wound around a soft iron core. By doubling the number of turns of wire on the secondary coil, twice the voltage is produced. By increasing the turn ratio ten times, ten times the voltage is produced across the secondary coil. However, the amperage will be *one tenth* so the total power remains the same.

EXPERIMENTS

The following experiments are laid out in a progressive order to help you get a good basic background of knowledge in electrical work.

MAGNETISM

As you have learned, magnetism can exist in either temporary or permanent form. The temporary form is employed in most industrial uses, as it can be turned on and off at will. A knowledge of magnetism is essential to an understanding of the operation of motors, relays, generators, controls, starters, and ignition coils.

Tools and Equipment

Hacksaw blade; iron filings; two bar magnets; horseshoe magnet; soft iron nail; string; chalk; sheet of paper; small flat wood board.

Procedure

TO DEMONSTRATE THE DIFFERENCE BETWEEN TEMPORARY AND PERMANENT MAGNETS

NOTE: *Answer the following questions in your notebook or on scratch paper as directed by your instructor. Do not write in this book.*

1. Magnetize a hacksaw blade by stroking one end of it with one pole of a strong bar magnet.
2. Dip the magnetized blade into the box of iron filings. What happens?
3. Tap the blade several times on a steel object, and then dip it into the iron filings. What happens?
4. Explain the different results according to the molecular theory.
5. Dip the tip of the soft iron nail into the filings. What happens?
6. Hold the tip of the nail about 1/8 inch over the filings. Bring one pole of a strong bar magnet close to, *but not touching,* the head of the nail. Hold a piece of cardboard between the two to keep the spacing. What happens?

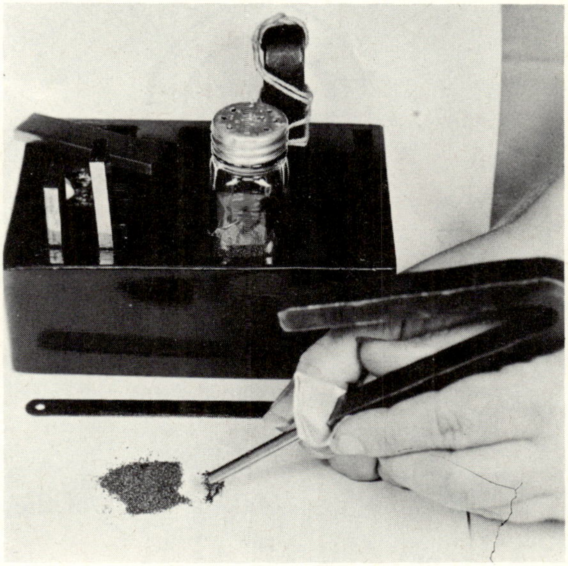

Magnetic induction. Note the folded paper to keep the spacing between the magnet and nail.

7. Without moving the nail, remove the magnet. What happens?
8. Explain magnetic induction.
9. In your own words, explain the difference between a temporary and a permanent magnet.

TO DETERMINE THE SHAPE OF THE FIELD OF FORCE ABOUT A MAGNET

1. Place a horseshoe magnet on the flat wood board. Cover the magnet with a sheet of paper.
2. Sprinkle iron filings on the paper. In your notebook, make a pencil sketch of the field of force which is revealed.
3. Place two bar magnets on the wood, with the north pole of one facing the south pole of the other, about ½ inch apart.
4. Cover with paper and sprinkle

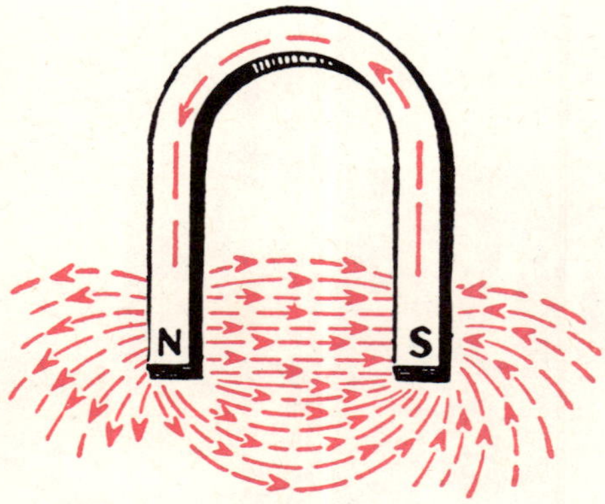

A field of force exists between the poles of a magnet.

with iron filings. Make a sketch of the field.

5. Place two bar magnets so that the north poles face each other.

6. Cover with paper and sprinkle with iron filings as before. Make a sketch of the field of force.

To Demonstrate the Law of Magnetic Attraction

1. Hang a bar magnet on a string so that it swings freely. It will come to rest in a north-south position. The end pointing to the North Pole of the earth is called the *north pole.* Whiten this end with a piece of chalk.

2. Bring one end of a second bar magnet toward the whitened (north pole) end of the hanging magnet, which either will be attracted or repelled.

The law of magnetic attraction states that like poles repel each other.

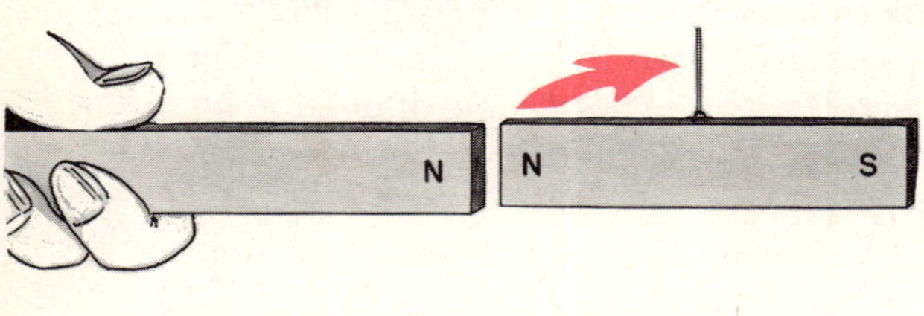

- If the hanging magnet is repelled, whiten the end of the magnet (held in your hand) which is closest to the north pole of the hanging magnet.
- If the hanging magnet is attracted, whiten the end of the magnet which is farthest from the north pole of the hanging magnet.
- Now, both magnets have their north poles whitened and the south poles unmarked.

3. Bring the north pole of the magnet toward the north pole of the hanging magnet. What happens?

4. Bring the south pole of the magnet toward the south pole of the hanging magnet. What happens?

5. Bring the north pole of the magnet toward the south pole of the hanging magnet. What happens?

6. Bring the south pole of the magnet toward the north pole of the hanging magnet. What happens?

7. Explain the law of magnetic attraction.

ELECTROMAGNETISM

In this section you will find that both permanent and electric magnets produce the same force. It requires three things to make an electric magnet: a coil of wire, a soft-iron core, and an electric current. By turning the electric current on and off, you will be able to control the magnetic force.

Tools and Equipment

Storage battery; ½ x 1½ inch bolt; ¼ x 1½ inch bolt; knife switch; O-10 ammeter; ½ inch brads; rheostat O-10 ohms; compass; bell wire; No. 26 SCC wire; cardboard with a hole in the center; solenoid of two-inch diameter form; flat soft steel, 9 x 1½ x 3/16 inch; welding rod, 1/8 x 6 inch; iron filings.

Procedure

To Demonstrate the Factors Affecting the Strength of an Electromagnet

1. Wind two coils around a piece of 5/8 inch dowel or rod, one of forty turns and the other of eighty, using No. 26 SCC magnet wire.
2. Insert the large bolt into the forty-turn coil and connect it to a storage battery in series with an ammeter, rheostat, and switch.
3. Adjust the rheostat until the ammeter indicates a flow of two amperes.
4. Touch the end of the bolt to the pile of brads and record the number attracted. Open the switch while making notes, to minimize the drain on the battery.
5. Substitute the eighty-turn coil of wire for the smaller one.
6. Close the switch and adjust the rheostat until the ammeter reads two amperes.
7. Touch the bolt to the pile of brads and record the number attracted.
8. Is the magnet weaker or stronger when the number of turns in the coil is increased?
9. Substitute a small bolt for the large one.
10. Readjust the rheostat until the ammeter reads two amperes.
11. Touch the bolt to the pile of brads and record the number attracted.
12. When the core size is increased, does the magnet become weaker or stronger?
13. Adjust the ammeter reading to one ampere and touch the bolt to the pile of brads.
14. Record the number attracted.
15. When the current flow is increased, does the magnet become weaker or stronger?
16. List the three factors which affect the strength of an electromagnet.

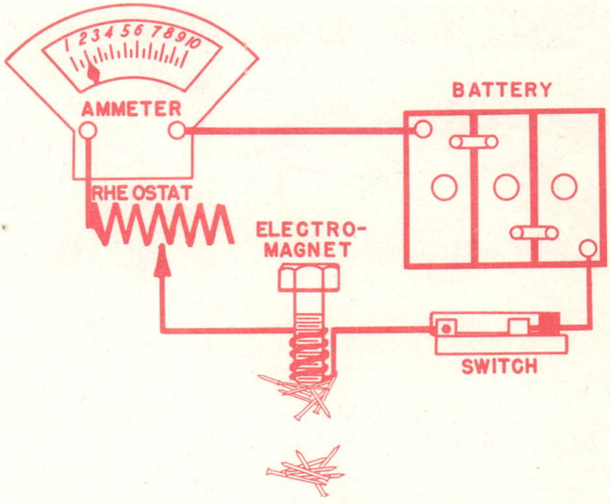

The strength of an electromagnet is dependent on the number of turns of wire, size of the core, and the amount of current flowing through the coil.

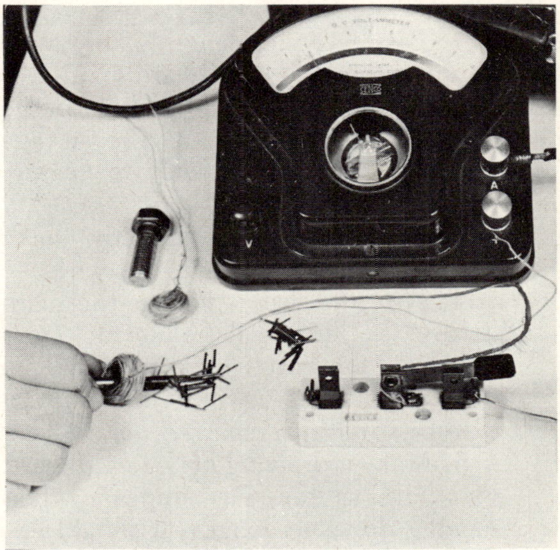

This is the equipment needed to show the difference between the strength of several types of electromagnets.

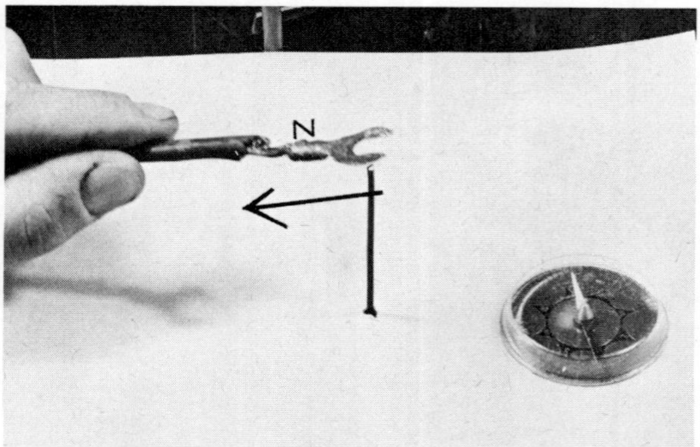

A circular field of force exists around a wire through which electricity flows.

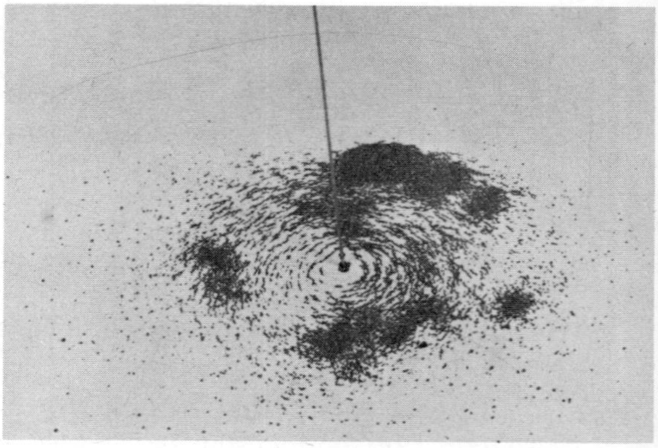

Sprinkling iron filings around a wire through which electricity is flowing indicates the circular field.

To Study the Fields of Force About a Wire, a Solenoid, and an Electromagnet

Wire

1. Pass a 15-inch length of bell wire through the hole in the cardboard.

2. Connect the bottom end of the bell wire to one post of the storage battery. **Note:** *Do not connect the wire to the other post of the battery.*

3. Place a compass directly north and 3 inches away from the wire.

4. *Flip* the free end of the wire across the first cell connector (two volts) of the battery. **CAUTION:** *Do not connect the wire to the battery post or it will become red hot.*

5. Make a drawing of the cardboard and wire as the wire appears when looking down its length—it should appear as a dot in the center of the cardboard.

6. Draw an arrow on your paper, pointing in the same direction as the compass needle when electricity flows through the wire.

7. Place the compass to the east of the wire and repeat the experiment and drawing.

8. Place the compass to the south of the wire and repeat.

9. Place the compass to the west of the wire and repeat.

10. Grasp the wire with your left hand so that your fingers circle the wire in the direction which the four arrows point. Note how your thumb indicates the direction in which electrons flow in the wire (negative to positive).

11. In your own words, explain the left-hand rule.

12. Sprinkle iron filings on the cardboard around the wire.

13. *Flip* the free end of the wire across the battery post.

14. Make a diagram of the field of force about the wire.

Solenoid

15. Connect the left end of the solenoid to the positive terminal of the storage battery. **NOTE:** *Do not connect the other end.*

16. Grasp the solenoid with your left hand so that your fingers follow the electron flow in the wires (negative

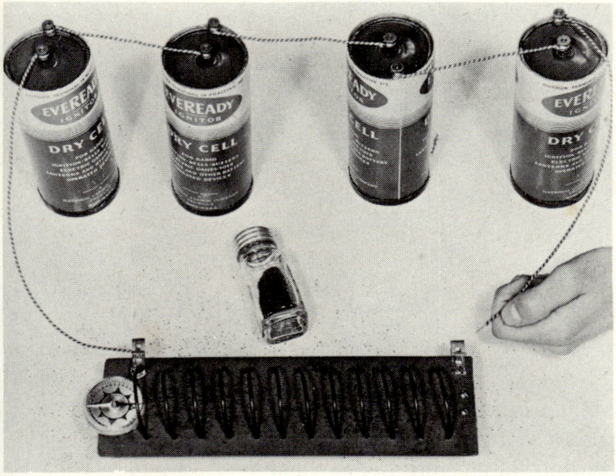

A solenoid exhibits the same characteristics as an electromagnet.

to positive). Your thumb will point to the north pole.

17. Place a compass about 8 inches from the end of the solenoid, which your thumb indicated to be the north pole.

18. *Flip* the free end of the solenoid connecting wire across the negative post of the battery. If your prediction of the north pole was accurate, the south pole of the compass needle will be attracted to the solenoid's north pole.

19. Reverse the battery connections to the solenoid and again predict the north pole with the left-hand rule. Flip the wire across the free battery post to check your prediction.

20. Sprinkle iron filings on the solenoid and flip the other wire across the free battery post. Sketch the resulting field of force.

ELECTROMAGNET

21. Place a flat piece of iron inside the solenoid to make it an electromagnet.

22. Predict the north pole with the left-hand rule.

23. Place the compass about 8 inches away from the predicted north pole and flip the other wire across the free battery post to check your prediction.

24. Has the north pole been changed by the insertion of the iron core?

25. Insert a piece of cardboard, cut to fit over the iron core, inside the solenoid.

26. Sprinkle iron filings on the cardboard and flip the other wire across the free battery post.

27. In what two ways has the field of force been changed from that produced by the solenoid?

28. Remove the iron core and substitute a piece of welding rod, laying it loosely in one end of the solenoid.

29. Flip the other wire across the free battery post to energize the solenoid.

30. What happens to the welding rod?

31. Push the rod to the other end of the solenoid and repeat.

32. Explain the mechanical action of a solenoid.

MEASURING ELECTRICITY

As previously mentioned, the *amount* or quantity of electricity flowing in a circuit is measured in amperes; electrical *pressure* is measured in volts. It is this pressure which forces current to flow from one point to another. The flow is hindered by *resistance;* the greater the resistance, the less electricity flows. The greater the pressure, the more flows.

Tools and Equipment

Four dry cells; bell wire; O-10 ammeter; O-10 volt meter; two-filament

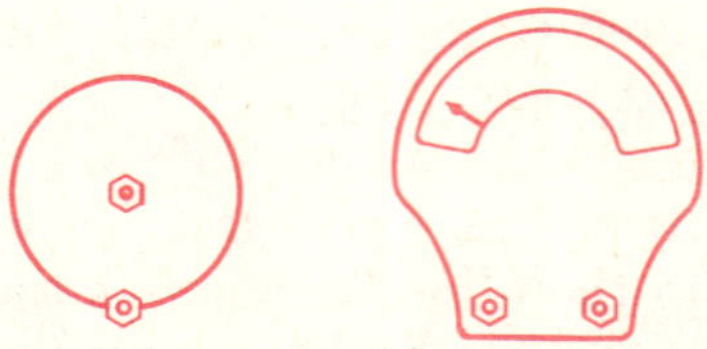

Measuring the voltage of one dry cell. On a piece of scratch paper, make a sketch of the battery, meter, and the wiring you are going to make. Have this paper checked by your instructor before you wire the circuit. *Do not make any pencil marks in this book.*

auto light bulb (32-32 CP); adjustable rheostat O-10 ohms; six-volt storage battery.

Procedure

VOLTAGE

1. Measure the voltage of one dry cell. **Note:** *If the meter reads off scale, reverse the connections.*
2. Hook up two dry cells in *series* and measure the voltage. Record.
3. Hook up two dry cells in *parallel.* Record the voltage reading.
4. Hook up four dry cells in *series parallel* and record. : *Hook up*

Measuring the voltage of two dry cells in series. In pencil, on scratch paper, make the necessary connections before connecting the actual wiring.

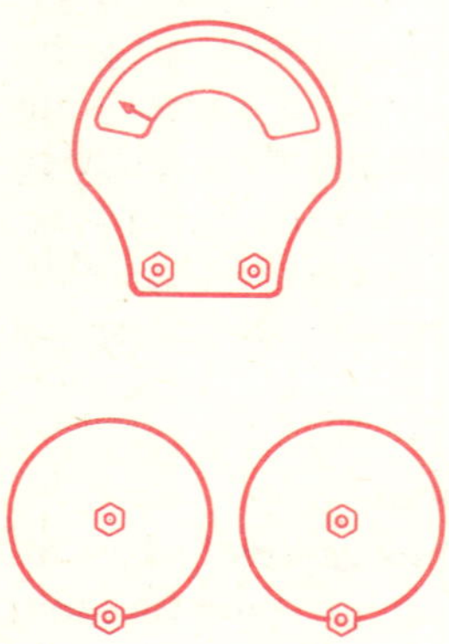

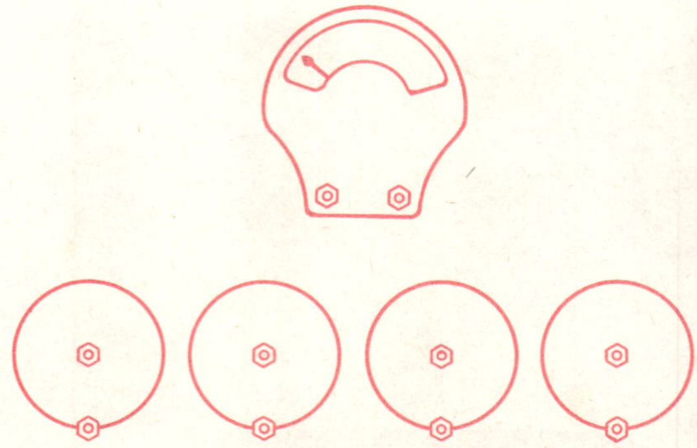

Measuring the voltage of four dry cells in series parallel. Pencil the connections on scratch paper before you make the actual wiring connections.

two dry cells in series; hook up the other two dry cells in series, and then connect both series sets in parallel.

AMPERAGE

1. Hook up two dry cells in series. Connect this voltage source in series with an ammeter and one filament of a bulb. Record the amperage.
2. Hook up four dry cells in series. Connect this voltage source in series with an ammeter and one filament of the bulb. Record.
3. When you increase the voltage, do you increase or decrease the amperage flow?

RESISTANCE

1. Hook up four dry cells in series.

Measuring the amperage of a headlight bulb under three volts pressure. Pencil the connections before you make the actual wiring connections.

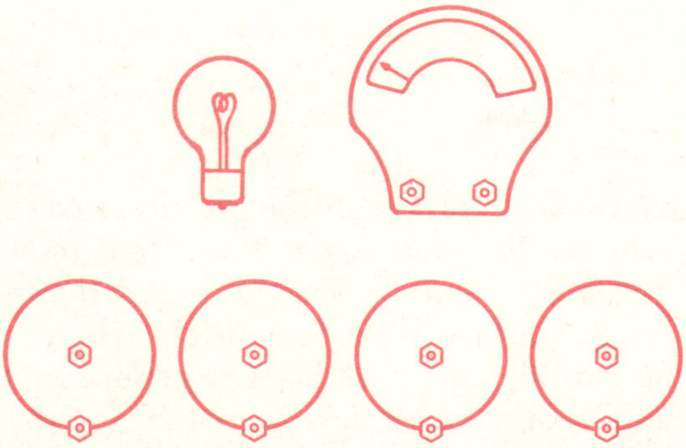

Measuring the amperage of a headlight bulb under six volts pressure. Pencil the connections before you make the actual wiring connections.

Measuring the amperage of two headlight bulbs in parallel under six volts pressure. Pencil the connections before you make the actual wiring connections.

Connect this voltage source in series with an ammeter, one filament of the bulb, and an adjustable rheostat.

2. Move the rheostat control arm to the right (maximum resistance) as far as it will go. Record the ammeter reading. How bright is the lamp?

3. Move the rheostat control arm to the center of the rheostat. Record the ammeter reading. Is the lamp bright, dim, or out?

4. Move the rheostat control arm to the extreme left (no resistance). Record the ammeter reading. Is the lamp bright, dim, or out?

5. When you increase the resistance,

Connecting a variable resistance in series with a headlight bulb. The variable resistance regulates the flow of current to the light. Variable resistances are frequently used to control the speed of the heater motor and the intensity of the dash lights. Pencil the connections on scratch paper before you make the actual wiring connections.

A handy wiring board to hold the headlamp wiring equipment. Note the diagram below the bulb mounting which shows how the filaments of the bulb are connected internally.

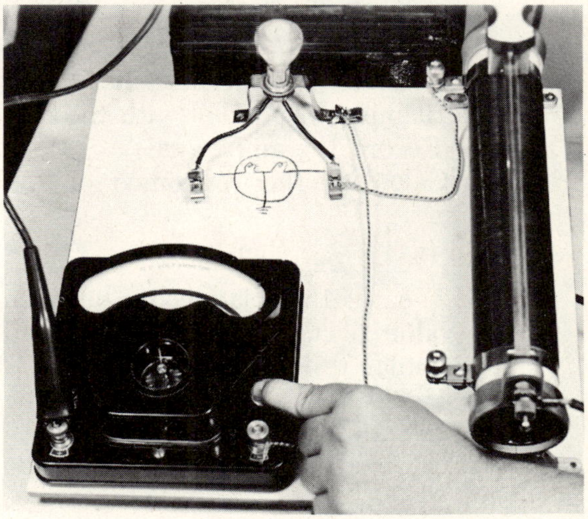

do you increase or decrease the amperage flow?

6. When does the lamp become brighter—when you increase or decrease the amperage?

7. When does the lamp become brighter—when you increase or decrease the resistance?

Ohm's Law

1. Connect a six-volt storage battery in series with an ammeter and one filament of a bulb. Record the amperage.

2. To calculate the resistance of the lamp bulb, substitute the amperage reading in the following formula:

$$R \text{ (resistance)} = \frac{E \text{ (voltage)}}{I \text{ (amperage)}}$$

$$\text{as } R = \frac{E}{(\quad)}$$

3. Measure the voltage of the storage battery you are using when the lamp is lit.

4. Substitute the value of "E" above in the formula:

R = ________.

5. Divide the value of "I" into the value of "E" in order to get the value of "R." As:

$$I\overline{)\,E\,} \quad \text{thus: } R = \underline{\qquad\qquad} \text{ ohms}$$
$$R$$

6. Connect the second headlamp bulb filament in parallel with the first one. Record the amperage.

7. Calculate the resistance of the circuit.

R = ________ ohms.

8. What is the relationship between the value of the first resistance and two similar resistances in parallel?

9. Hook up both filaments of the bulb in series. Record the amperage.

10. Record the voltage when the bulb is lit.

11. Calculate the resistance of two filaments in series.

R = ________ ohms.

NOTE: *The resistance of the two filaments in series is not twice that of the single filament.* This is due to the fact that the bulb is not as bright when the two filaments are hooked in series. Because the resistance increases with an increase in temperature, you would expect to find proportionally less resistance with the two filaments hooked in series because of the dimness of the bulb.

12. Is the resistance greater in a parallel or series circuit?

13. Explain step 12 in your own words.

WIRING ELECTRICAL CIRCUITS

Electricity may be conducted to bells, lamps, heating devices, and motors by means of wires. These wires must be connected properly for the unit to work. In many cases, switches are installed in the circuit to control the flow of electricity.

Tools and Equipment

Two bells; one buzzer; one button; two single-pole knife switches; one double-pole, double-throw knife switch; one reversing switch; wire; four dry cells; screwdriver; pliers.

Procedure

1. Hook up four dry cells in series as a voltage source for the following experiments:

2. Study and apply the following three rules of wiring:

• a. Connect one pole of the battery to one pole of the bell or group of bells.

• b. Connect the other pole of the bat-

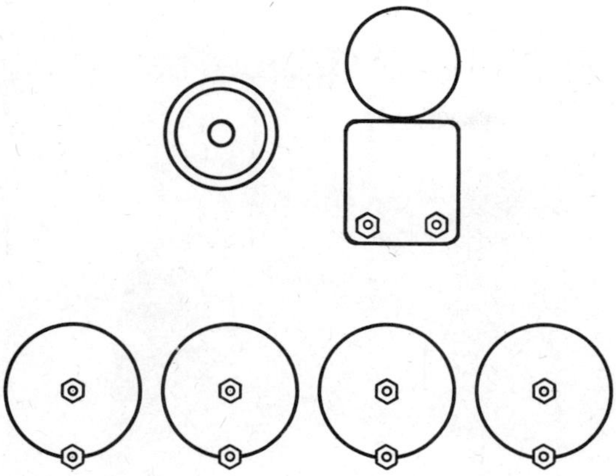

One button and one bell. This is a basic circuit used on all automotive electrical equipment such as ignition, starter, lights, and radio. Pencil the connections on scratch paper before you make the actual wiring connections.

tery to one pole of the switch or switches.

• c. Connect the remaining pole of the switch to the remaining pole of the bell it is to control.

3. Connect one button to ring one bell. This type of connection is similar to that used to operate the horn.

Two buttons and one bell. This switching arrangement is used to light the interior of a car when either door is opened. Pencil the connections on scratch paper before you make the actual wiring connections.

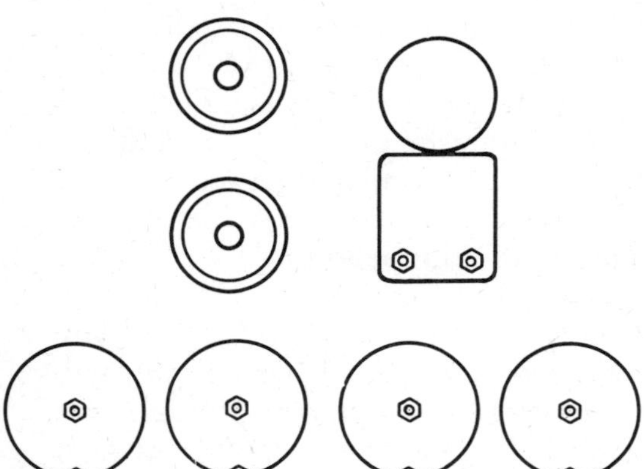

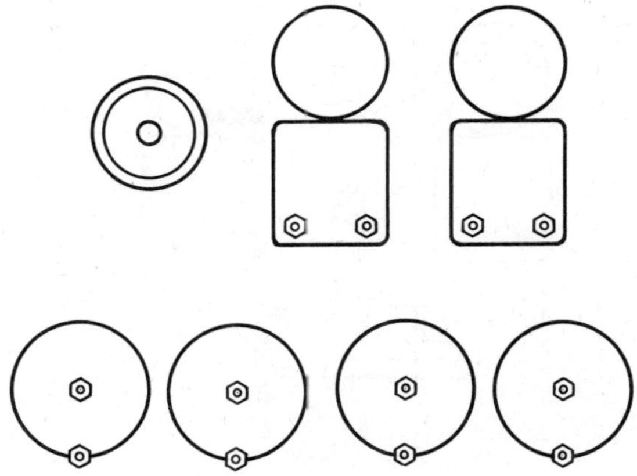

One button and two bells. This electrical circuit is basic when wiring any two lights or horns to a switch. Pencil the connections before you make the actual wiring connections.

4. Connect two buttons to ring one bell. This kind of circuit is used to turn on the interior body light of a car. Either car door will operate it.

5. Connect one button to ring two bells. This kind of circuit is used to connect the headlamps and tail lamps to a light switch.

Two buttons, one bell, and one buzzer. This double switch circuit is an exact duplicate of the direction signaling lights on the car. Pencil the connections on scratch paper before you make the actual wiring connections.

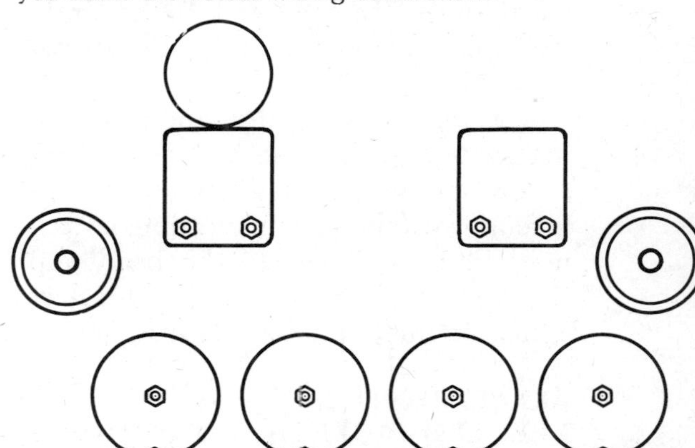

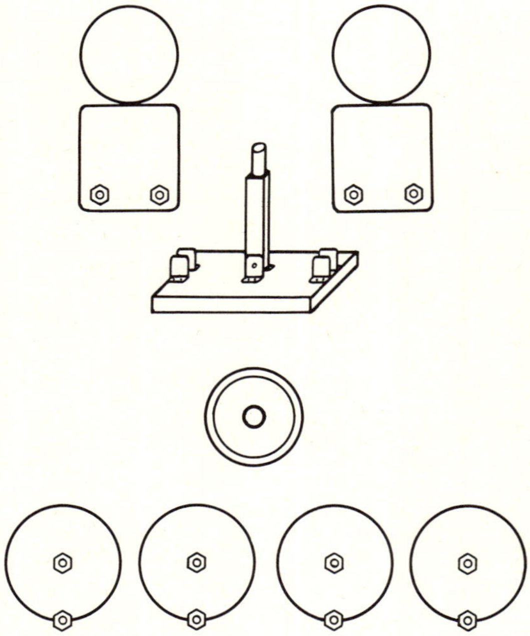

Two bells, one button, and one reversing switch. This reversing switch circuit is an exact duplicate of the high and low beam selector switch (tilt-ray) used on the car. Pencil the connections before you make the actual wiring connections.

The picture shows the construction of a board which will simplify the wiring of all bell experiments.

6. Connect two buttons, one bell, and one buzzer so that one button will ring the bell and the other button will operate the buzzer. This type of connection is used whenever any two electrical units are operated by separate switches, such as the dash and interior lights of a car.

7. Connect one button and a reversing switch so that the button controls either of two bells selected by the reversing switch. The reversing switch used here is similar to the headlamp tilt-ray switch, which is used to select either the high or low bright lights.

EXPERIMENTS IN ELECTROMAGNETIC INDUCTION

Michael Faraday's discovery of the principle of electromagnetic induction in 1831 led to his producing electricity from magnetism. His success made possible the development of the generator. As previously mentioned, he noted that movement of a coil of wire, through a magnetic field of force, created enough electricity in the coil to move a galvanometer needle.

In this job you will study the principles of electromagnetic induction to aid in understanding the principles of the generator and induction coil.

Tools and Equipment

No. 10 copper wire, 3 feet long; twenty-turn coil of No. 10 copper wire wrapped about a 1-inch form; galvanometer; bar magnet; storage battery; bell wire; No. 20 SCC magnet wire; laminations from an old radio power transformer; three 1-inch coil forms; two AC voltmeters 0-15 volts; bell-ringing transformer; ignition board; horseshoe magnet.

Procedure

To Demonstrate the Effects of a Coil of Wire Being Passed Through a Magnetic Field

1. Connect the ends of the 3-foot length of copper wire to a galvanometer.
2. Pass the loop of wire quickly between the poles of the horseshoe magnet, and then hold it stationary.
3. What happens to the galvanometer needle?
4. What has happened within the wire?
5. Lift the wire quickly between the poles of the magnet, and then hold it still.
6. What happened to the galvanometer needle?
7. Can you explain why the needle moved in a different direction in steps 3 and 6?
8. Connect the twenty-turn coil of No. 10 wire to the galvanometer. Remove the coil form so that the coil is air wound. It will support itself because of the size of the wire.
9. Thrust one end of a bar magnet into the coil and hold it motionless after it reaches the center.
10. What happens to the galvanometer needle?
11. Why does the needle return to zero after the magnet stops moving?
12. Quickly pull the magnet from the coil.
13. What happens to the galvanometer needle?
14. Hold the bar magnet stationary and move the coil back and forth over the magnet. Is there any difference between movement of the coil and movement of the magnet?
15. What is necessary to produce a constant flow of electricity?

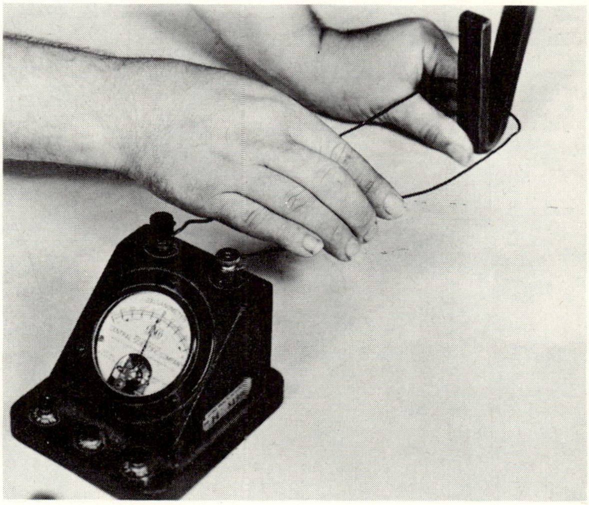

Electromagnetic induction. The wire must be moved quickly across the magnetic lines of force to produce electricity.

To Demonstrate the Properties of a Transformer

1. Remove the laminations from an old radio power supply transformer. You will find them to be shaped like an "E."
2. Cut out the center section of the "E" so that it becomes "C" shaped.
3. Mount the laminations on a board with the two ends of the "C" pointed up so that you can place different coils over the ends for study.
4. Cut a piece of soft steel to cover the top gap so that the magnetic lines of force can pass completely around.
5. For study, wind three coil forms with the No. 20 SCC wire. Wind two each of 200 turns and one of 400 turns.
6. Connect a bell-ringing trans-

The transformer. While the efficiency of this unit is low, the ability to substitute coils of different sizes, while observng the voltage changes, makes it a very efficient learning tool.

former to the 110 volt AC line in order to reduce the working voltage to about 10 volts. This is a safety measure to prevent you from getting a strong shock.

7. Place the two 200-turn coils over the two upright ends of the laminations and install the top gap bar.

8. Connect one coil to the low voltage side (primary) of the transformer.

9. Connect the other coil (secondary) to the resistor load.

10. Connect a voltmeter across each coil to observe the working voltages.

11. Plug in the 110 volt AC line.

12. What is the voltage of the primary coil?

13. What is the voltage of the secondary coil?

14. The difference in voltage between the primary and secondary is due to the inefficiency of the transformer and not to any step-down ratio. To prove this, interchange the two coils and the voltages will remain constant.

15. How do you account for electricity being produced in the secondary coil even though there is no movement of either coil?

16. Hold a thin strip of steel such as a hacksaw blade close to the primary coil laminations. What happens to the blade?

17. What does this indicate is happening to the primary coil?

18. What causes the vibration? Is there a steady flow of magnetism, or is the flow pulsating?

19. Can you now account for the magnetic movement from an AC coil as compared with the mechanical movement of a bar magnet?

20. Substitute the 400-turn coil for the secondary and record the secondary voltage. This is a step-up transformer.

21. What is the ratio of the primary to secondary turns?

22. What is the difference between the voltages you found in the secondary in steps 13 and 20?

23. Change the primary and secondary coils so that you have a step-down transformer.

24. Measure the voltage of the secondary.

25. What is the ratio of the primary to secondary turns?

26. What is the ratio of the voltages in the secondary in steps 13 and 24?

TO DEMONSTRATE THE PROPERTIES OF AN INDUCTION COIL

1. Hook up the induction coil board as shown. Use the schematic symbols painted on the board to assist you in making connections.

The induction coil. The wiring diagram painted on the board will assist you in making the proper connections.

2. Connect the two input terminals to a storage battery.

3. Adjust the high tension spark gap so that a spark occurs as you turn the cam which opens and closes the points.

4. Watch the points carefully and turn the cam to open and close the points just once.

5. Does the spark occur when the points open or when they close? The ignition coil is a step-up transformer, changing the voltage of the storage battery to about 15,000 volts, which is needed to jump the gap.

6. Disconnect the condenser wire from the circuit.

7. Open and close the points.

8. Do you get a high-voltage spark?

9. Adjust the high-tension spark gap closer until you do get a spark.

10. Turn the cam to open and close the points.

11. Watch the points carefully to note the arcing and burning which occur without a condenser.

12. Connect the condenser back in the circuit and repeat. Note the absence of the arcing and burning.

13. What effect has the condenser on the action of the points?

14. What effect has the condenser on the output of the coil?

CHECK YOUR KNOWLEDGE

1. Why is magnetism considered the basis of electricity?
2. What is loadstone?
3. What is the present theory of magnetism?
4. What is the difference between a permanent and temporary magnet?
5. Of what three things is a battery made?
6. What is an open circuit?
7. What is voltage?

8. What is amperage?
9. What is a complete circuit? Series circuit? Parallel circuit?
10. Explain how an automotive 12-volt battery is an example of series and parallel connections.
11. Describe an electromagnet.
12. What are ampere-turns?
13. What is the left-hand rule?
14. What is a solenoid? How does it work? Of what use is it?
15. What is a relay? How does it work? Of what use is it?
16. Why does it take three switches to operate the starting motor?
17. What is a circuit breaker? How does it work? Of what use is it?
18. What is resistance? How is it measured?
19. What is Ohm's law? Of what use is it?
20. What is induction?
21. On what three things does the strength of an induced current depend?
22. What is mutual induction? Self-induction?
23. What is the purpose of the primary coil of a transformer? Secondary coil?
24. What action does varying the turns ratio of a transformer have on its output voltage?

Self-Check Tests

TRUE-FALSE

1. Permanent magnets are made of soft steel.
2. The regions around the ends of a magnet are called the poles.
3. The law of magnetic attraction states that like poles repel each other.
4. The south pole of a magnet points toward the geographic North Pole.
5. A large dry cell contains more voltage than a small one.
6. A large dry cell contains more amperage than a small one.
7. Battery cells are connected in series to increase the total amperage.
8. The polarity of an electromagnet can be predicted by the right-hand rule.
9. A relay has a movable soft-iron core.
10. A circuit breaker is a form of solenoid.
11. An ohm is the amount of amperage which flows with a pressure of one volt.
12. Ohm's law states that the resistance is divided by the voltage to calculate the number of ohms in the circuit.
13. A galvanometer is necessary for induction.

MULTIPLE CHOICE

1. The first natural magnet was called: a) loadstone; b) temporary magnet; c) permanent magnet.
2. The difference in electrical pressure between two poles of a battery is called: a) amperes; b) ohms; c) volts.
3. A large dry cell furnishes more: a) amperes; b) ohms; c) volts than a small one.
4. A circuit in which all of the electrical energy must pass through all of the electrical devices is called a: a) series; b) parallel; c) open circuit.
5. A circuit which has many branches for electrical energy to pass is called a: a) series; b) parallel; c) open circuit.
6. Batteries are frequently connected in series to increase the: a) amperage; b) voltage; c) flux.
7. An electromagnet containing a soft-iron core is called a: a) relay; b) armature; c) solenoid.
8. A circuit breaker takes the place of: a) switch; b) fuse; c) open circuit.
9. The amount of resistance of a wire does *not* depend on the: a) size; b) length; c) temperature; d) cost.
10. Ohm's law states that the resistance of a circuit can be calculated by: a) voltage divided by the amperes; b) amperes times the voltage; c) amperes divided by the voltage.
11. The amount of electricity induced in a wire is *not* dependent on which of the following: a) number of turns of wire; b) strength of the magnetic field; c) speed of movement; d) size of wire?
12. By doubling the number of turns of

wire of a transformer primary coil: a) one half; b) the same; c) twice the voltage will appear across the secondary coil.

COMPLETION

1. An ________ ________ can produce magnetism.
2. The first natural magnet was called ________.
3. A magnet made of ________ steel is called a temporary magnet.
4. A ________ is a container of acid in which two dissimilar metals are submerged.
5. The electrical p r e s s u r e of a fully charged wet cell is always ________ volts.
6. The quantity of electrons flowing in a circuit is called ________.
7. The voltage of batteries hooked in ________ remains the same as for one battery.
8. The strength of an electromagnet is classified according to the number of ________ ________.
9. A ________ is an electromagnet with a movable soft-iron core.
10. A relay is an electromagnet with a hinged ________.
11. A circuit breaker takes the place of a ________.
12. Resistance is electrical friction and is measured in ________.
13. The process by which electric current is set in motion in a wire through movement in a magnetic field is called ________.
14. The ________ coil of a transformer is the one from which current is taken.

MATCHING

1. Loadstone	1. Region around poles
2. Soft steel magnet	2. Ampere turns
3. Hard steel magnet	3. Wet cell
4. Ends of a magnet	4. Temporary magnet
5. Field of force	5. Induction
6. Potential difference	6. Movable soft-iron core
7. Quantity of electricity	7. Poles
8. Storage battery	8. Permanent magnet
9. Electromagnet	9. Self-induction
10. Solenoid	10. Natural magnet
11. Circuit breaker	11. Amperage
12. Resistance	12. Ohm
13. Transformer	13. Takes the place of a fuse
14. Mutual induction	14. Voltage

CHAPTER 3
SMALL ENGINES

Small gasoline engines have come into widespread use, replacing muscle power for many tasks and thereby helping us to enjoy life. These engines are used mainly in boats, motor scooters, motorcycles, and for powering tools and equipment. In rural areas they are used to drive generators if electric power is not available.

Miniature Vehicles

For years, children have built small cars out of old wagons and soap boxes, but now through the progress of thought and technology these innocent toys have been transformed into machines with power called carts, karts, or go-carts, etc. Operation of these carts is accompanied with hazards, some of which are similar to those of operating full sized motor ve-

Parts of the go-kart. See exploded view, Page 73. Courtesy Go-Kart Manufacturing Company.

hicles, and others which seem to be peculiar to miniature vehicles.

The go-cart has wide popularity which has created strong desires for children of less than legal driving age to drive them. However, an important reason for drivers license laws is that below the stated age, children have attained neither the skill nor judgment to operate a motor vehicle. The importance of this subject was sufficient to cause the National Safety Council to issue a formal policy on miniature motor vehicles. This policy is as follows:

"The National Safety Council opposes the use of miniature motor vehicles, commonly called carts, or go-carts, by drivers below licensing age.

Cutaway view of a small engine. Courtesy Briggs & Stratton Corporation.

Exploded view of the chassis. 1. Frame assembly, 2. Spindle, 3. Spindle, 4. Tie rod, 5. Wheel and tire assembly, 6. Rear wheel and tire assembly, 7. Brake pedal, 8. Throttle pedal, 9. Rear axle shaft, 10. Bearing, 11. Collar, 12. Bearing carrier, 13. Bolts, 14. Nuts, 15. Lock washers, 16. Key, 17. Axle nut, 18. Sprocket, 19. Brake disc, 20. Brake rod, 21. Throttle rod, 22. Return spring, 23. Return spring washer, 24. Brake assembly, 25. Lock nut, 26. Fuel tank. Courtesy Go-Kart Manufacturing Company.

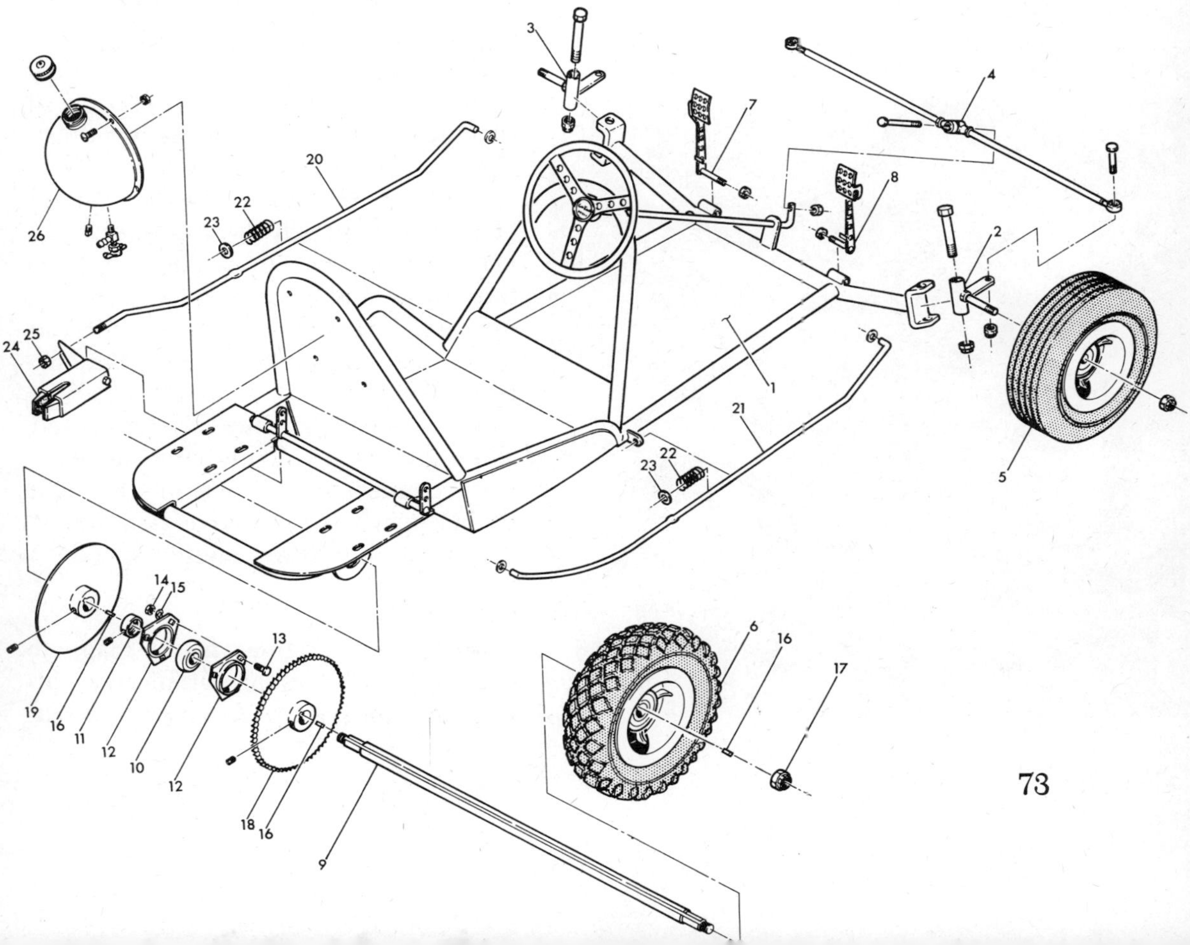

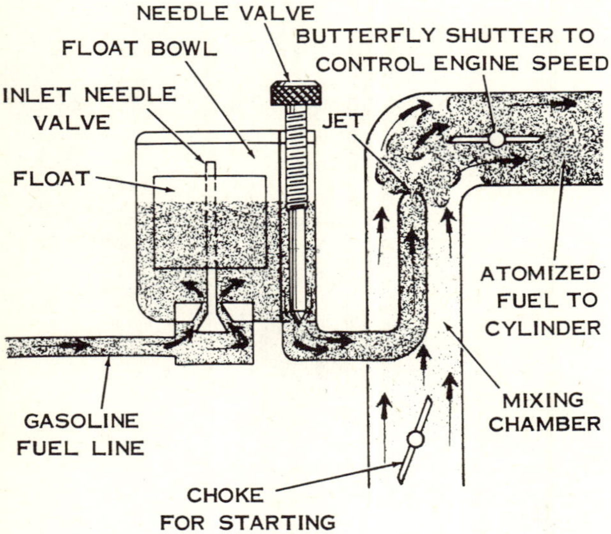

Small engine carburetors may contain only a single jet. Courtesy Lawn-Boy Division, Outboard Marine Corporation.

Some small engines have a primer in the form of a lever which operates the fuel pump to prime the system after long layovers. Courtesy Wisconsin Motor Corporation.

"While the laws of all states exclude these miniature vehicles, designed primarily for use by children, from the public streets and highways, there is little or no control over their use on private property. As a result, children have been killed or injured operating these vehicles, sometimes in speed events sponsored by adults. Motor vehicle speed events of any type are opposed by the National Safety Council.

"The Council therefore urges parents, dealers in these vehicles, and organizations to refrain from sponsoring or conducting competitive events using miniature motorized vehicles, and to do all possible to see to it that they are at no time operated by persons under legal driver age."

CARBURETION

Small engine carburetion is usually much simpler than automotive carburetion. Usually only one or two jets are used. Many engines run at a governed speed and consequently need only a single-jet carburetor to supply a fuel-air mixture. The proportion of fuel in the mixture is usually adjustable by a needle valve.

A float circuit maintains a reservoir of fuel from which the jets operate. A venturi is used to draw the fuel from the nozzle and to atomize the fuel-air mixture. Few small engine carburetors have an acceleration pump or automatic choke.

Many small engines are governed. The governor weights are mechanically connected to open or close the carburetor butterfly valve in order to maintain the desired engine speed.

IGNITION

Most small engines use a magneto in place of the usual automotive battery-ignition system. Absence of a bat-

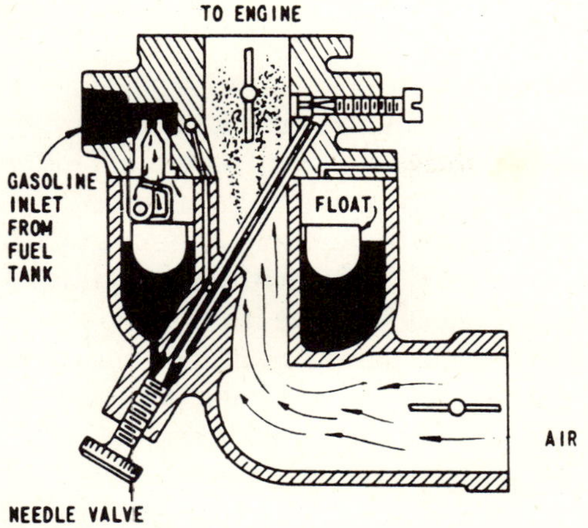

The float circuit maintains a reservoir of fuel for the high speed and idle circuits. Courtesy Briggs & Stratton Corporation.

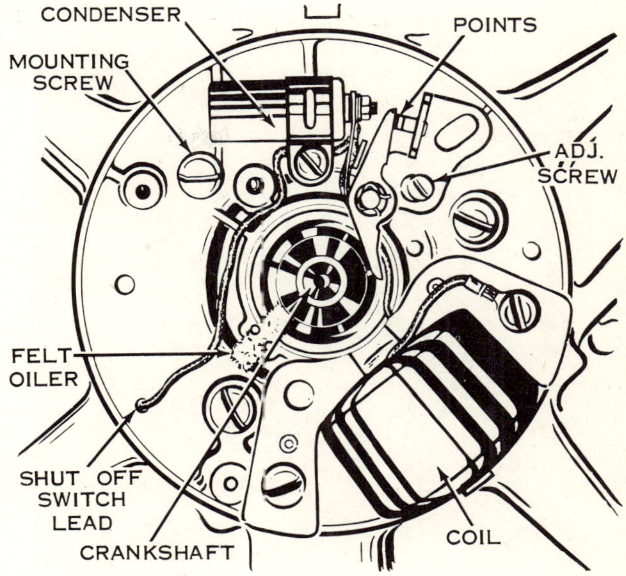

Parts of the magneto. Courtesy Lawn-Boy Division, Outboard Marine Corporation.

tery makes hand starting a necessity on such engines. The *magneto* is a self-energized electrical unit of the flywheel type which generates its own primary circuit current. The *coil, condenser,* and *breaker mechanism* are very similar to those used on the battery-type ignition system.

To generate its own primary current, the magneto is made of a primary coil and a permanent magnet, housed in the flywheel. As the flywheel rotates, the magnet moves past the primary coil and core, generating a voltage. When this voltage builds up to a maximum, the primary circuit breaker points open, causing the magnetic lines of force to collapse and induce a high-voltage surge across the secondary coil. The current is carried to the spark plug to ignite the fuel-air mixture.

FUELS AND LUBRICANTS

Most small engines require untreated fuel. Two-cycle engines tend to foul spark plugs when leaded fuels are used. This is also true of detergent-type oils. Consequently, it is good practice to use white gas, if it is available, and a motor oil ML or MM for light or medium duty without detergents or additives.

Gum is a product of leaded gasolines and, as small engines may be stored for long periods, gum formation is a

Schematic diagram of the magneto ignition system. Courtesy Briggs & Stratton Corporation.

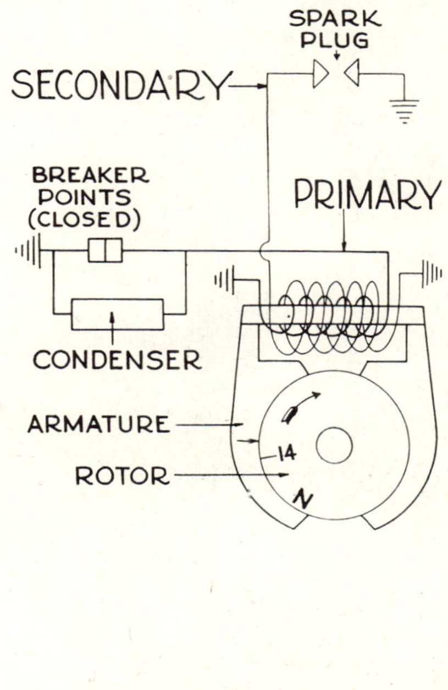

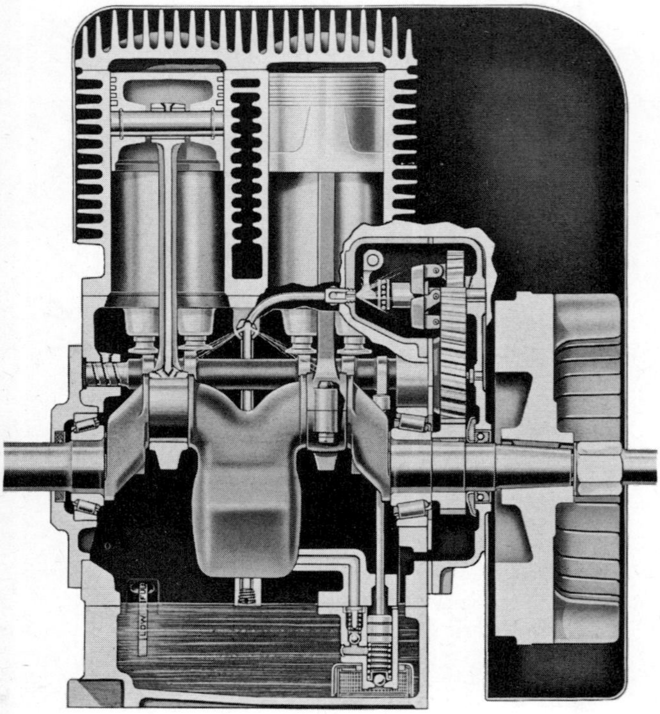

The oil pump, operated by a cam, sprays oil to all moving parts. Courtesy Wisconsin Motor Corporation.

Cross-section view showing the position of the pump and camshaft. Courtesy Wisconsin Motor Corporation.

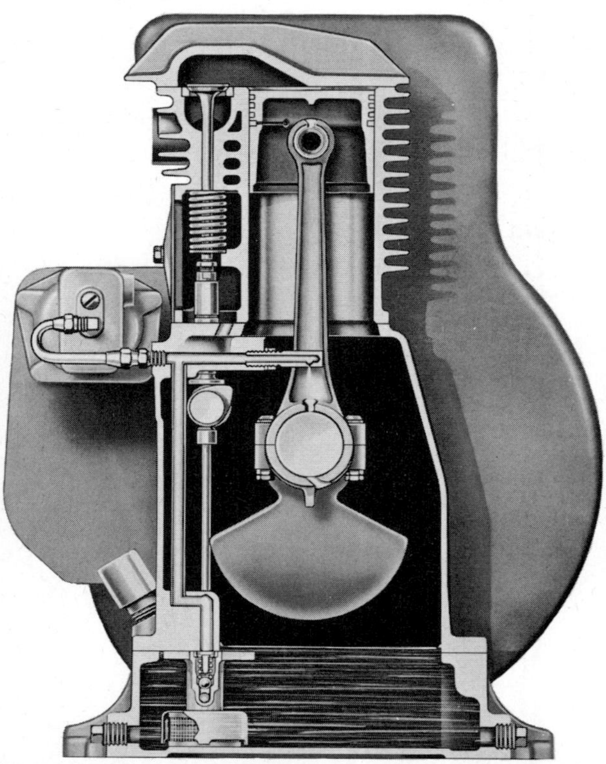

source of trouble. Always drain the gasoline from the tank and carburetor before storing a small engine. This is not necessary if white gas is used.

For two-stroke cycle engines, mix ½ pint of SAE 30 oil with each gallon of regular gasoline. The mixing should be done in a separate container and then poured into the gas tank in order to provide a uniform mixture.

SERVICE PROCEDURES

Most small engines work under extremely dusty conditions and generally require a complete overhaul at regular intervals, therefore the repair processes in this chapter are arranged about the complete overhaul of both one- and two-cylinder engines and of both two- and four-stroke cycle engines. To provide as broad a background as possible, one of each of the above types of engines will be disassembled.

Inasmuch as each one of these engines may be used in any one of several applications, the engines chosen for disassembly were selected on the basis of offering maximum variations in construction and application. In addition, three of the repair processes describe adjustments to commonly found, power-driven equipment: A motor scooter; a lawn mower; and an outboard motor. The fourth engine is of the type usually found in continuous service, providing power for such equipment as water pumps and electric generators.

Because the assembly instructions, with minor variations, are the same as the reversed disassembly instructions, they will not be repeated except for the variations. The general repair processes which apply to all engines will be found after the disassembly section.

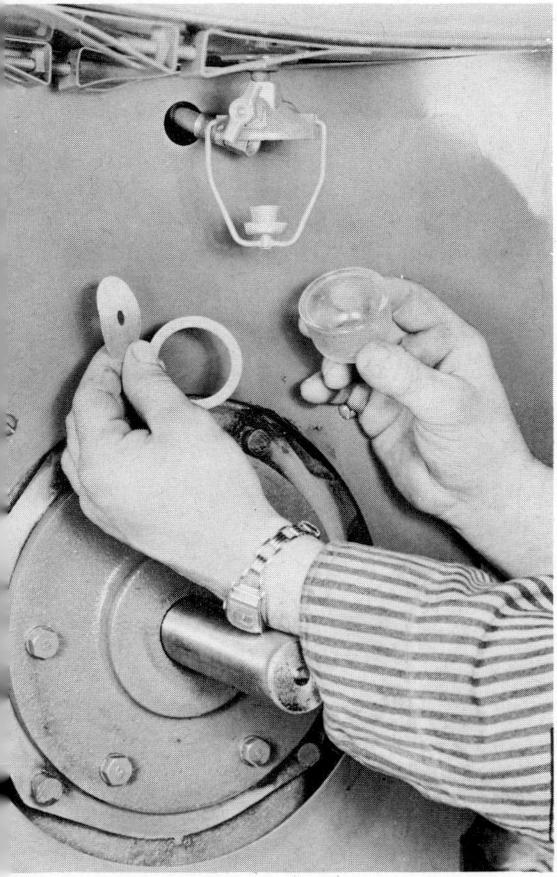

Cleaning the sediment bowl removes dirt and water before it can foul the carburetor. Courtesy Wisconsin Motor Corporation.

TROUBLESHOOTING

Troubleshooting the four-stroke cycle engine is similar to automobile work. The following problems and causes are especially characteristic of two-stroke engines but some of them also apply to four-stroke cycle engines.

Fuel System Troubles	Causes
1. Engine does not start	• Fuel tank empty • Water in fuel • Improper fuel mixture • Gummed up fuel line
2. Engine has no power	• Fuel mixture too rich or too lean
3. Engine stops	• Vent hole plugged • Fuel line clogged • No fuel in tank

Carburetor Troubles	Causes
4. Engine floods	• Choke stuck in closed position • Heavy float • Leaking needle and seat • Float level too high • Reed valve stuck open
5. Engine spits	• Water in gas • Dirt in jets • Gummed or plugged fuel line • Float level too low
6. Engine has no power	• Dirty jets • Improper mixture • Loose gasket sucking air • Reed valve stuck open • Clogged air cleaner

Spark Plug Troubles	Causes
7. Dirty black deposits	• Excessive idling • Too much oil in fuel mixture • Dirty breaker points • Too cool a spark plug
8. Pitted and burned	• Excessive high speed operation • Insufficient oil in fuel mixture • Broken fan • Too hot a spark plug

The ignition system can be tested by holding a wire close to any ground to see the size of the spark while cranking the engine. Courtesy Wisconsin Motor Corporation.

Magneto Troubles	Causes
9. Weak spark	• Dirty contact points • Points improperly adjusted • Weak condenser • Weak magnets • Broken or loose connection
10. Intermittent spark	• Broken or loose wire • Bare wire shorting on metal base • Broken point spring • Dirty or pitted points

Troubles encountered in the magneto ignition system. Courtesy Lawn-Boy Division, Outboard Marine Corporation.

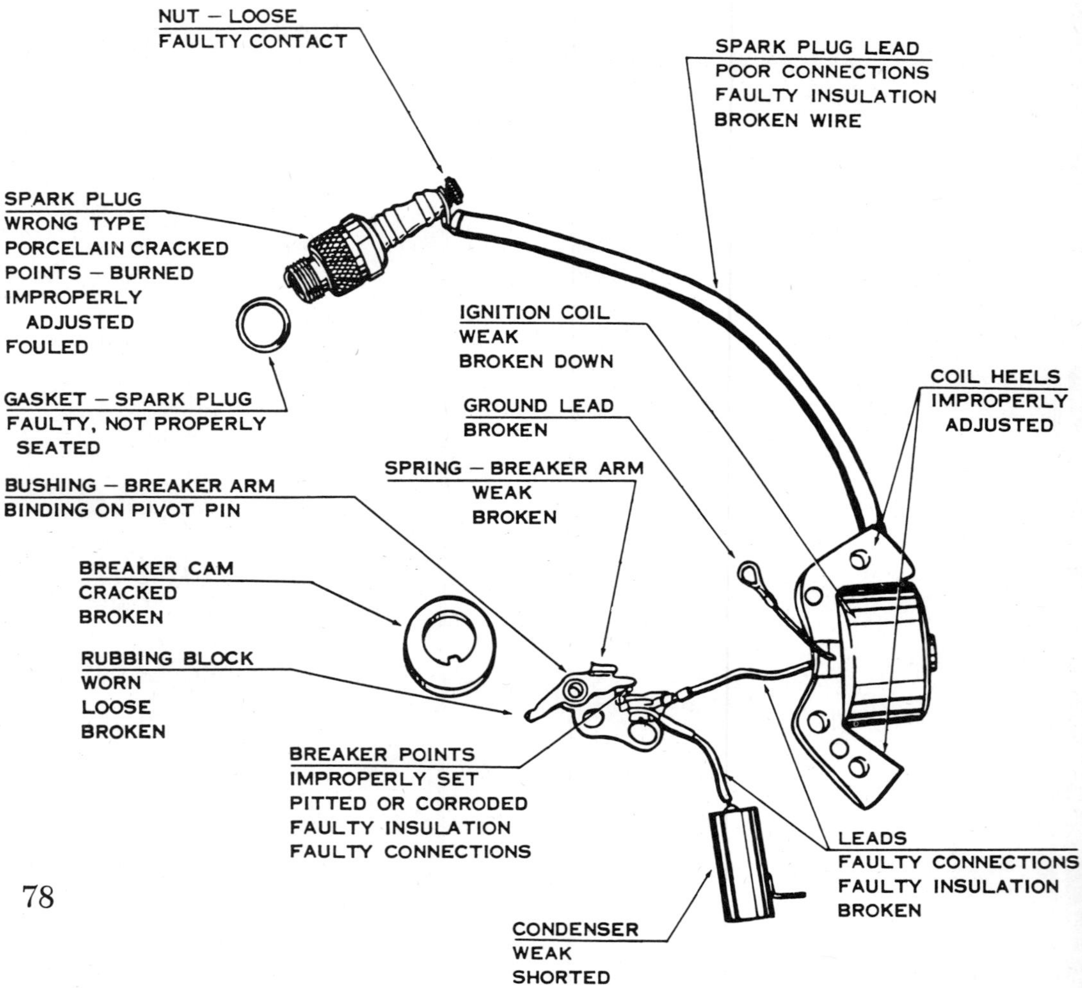

11. No spark	• Dirty contact points • Points not opening • Defective condenser • Defective coil • Shorted primary wire
Mechanical Troubles	**Causes**
12. Engine overheats	• Broken piston rings • Burned piston • Piston fitted too tightly • Improper lubrication • Crankshaft sprung • Fan loose or broken
13. No compression	• Stuck piston rings • Excessive blow-by • Worn cylinder walls • Defective cylinder head gasket • Inadequate lubrication • Carbonized ports
14. Noisy	• Excessive carbon • Piston slap • Worn piston pin • Loose bearings • Inadequate lubrication

GENERAL DISASSEMBLY INSTRUCTIONS (All engines)

When disassembling an engine, it is advisable to use several containers so that parts belonging to certain groups can be kept together. Capscrews of various lengths are used in the engine, therefore, care must be exercised during assembly so that, to prevent damage, the right screws are used in the right places.

Courtesy Lambretta Division, Innocenti Corporation.

DISASSEMBLING A TWO-CYCLE, ONE-CYLINDER ENGINE

(Each of the following steps is illustrated. The number in the lower right hand corner of the illustration agrees with the numbered steps below.)

1. First detach the side panels and the side boards.

2. Remove the spark plug and the high voltage cable. Close the gas cock and remove the carburetor. Then take out the baffle plate.

Courtesy Lambretta Division, Innocenti Corporation.

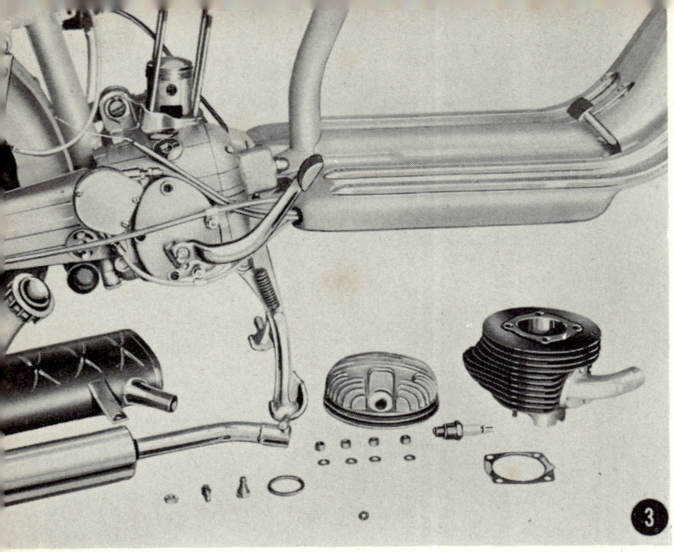

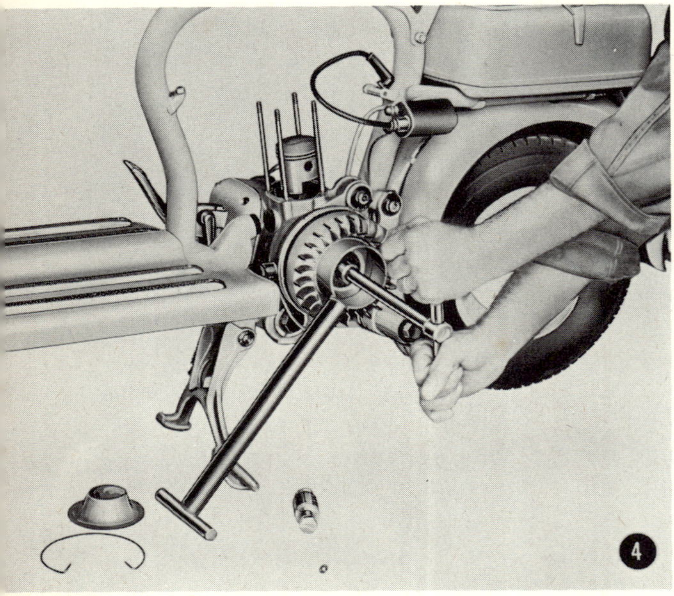

3. Unscrew the four cylinder-head nuts, and then remove the head. To remove the cylinder, loosen the ring between the cylinder and the exhaust pipe and unscrew the nut holding the muffler to the engine stud bolt. Free the muffler, after loosening the screws which hold it to the crankcase. Lift the cylinder from the studs and turn the crankshaft upward until the piston is entirely out of the case.

4. To remove the magneto, take off the cover which is held in place by a spring lock. Remove the rotor. Next unscrew the lock nut. Use a puller to free the stator from the crankpin. Disconnect the cables from the terminal. Remove the flywheel magneto locking key from its seat on the crankshaft. By loosening the three retaining screws, the core and coil can be removed.

5. When removing the crankshaft lock nut, insert a locking tool to prevent rotation.

6. Fasten a puller to the stationary flange and pull the flywheel. **CAUTION:** *Match the rotor and the stator plate immediately after removing to keep the pole pieces from losing their magnetism.*

Courtesy Lambretta Division, Innocenti Corporation.

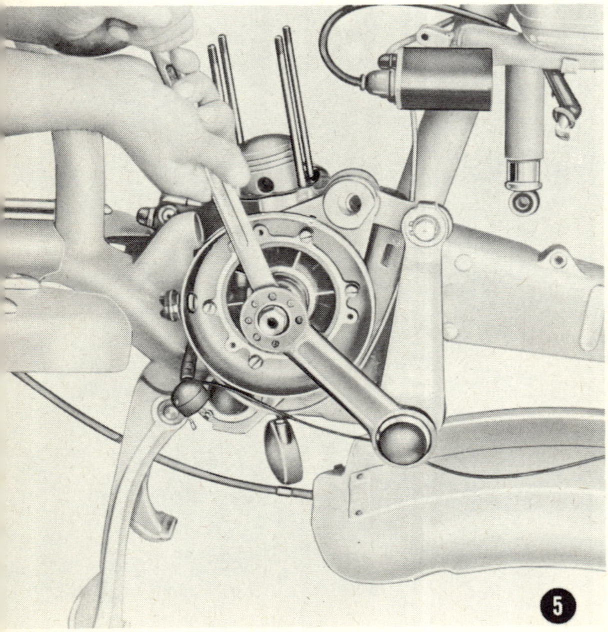

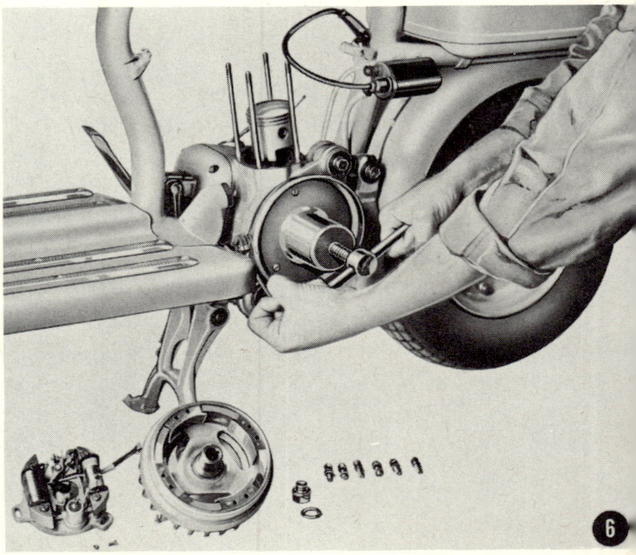

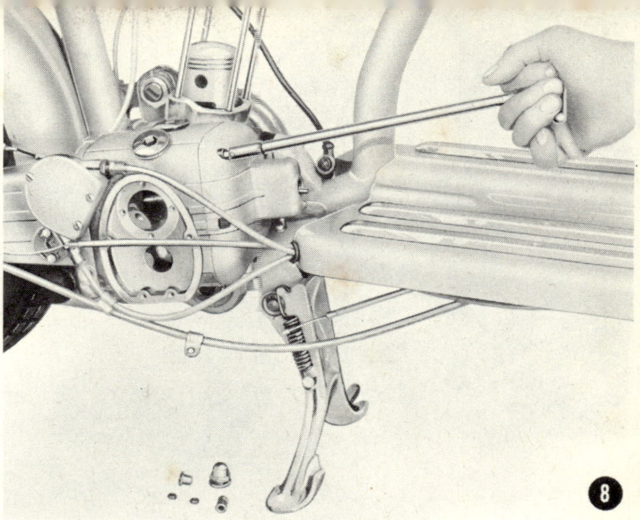

7. To remove the kick starter case, loosen the five bolts holding the cover to the crankcase. Loosen the two screws holding the bevel pinion support and remove this group.

8. To remove the clutch, disconnect the brake control cable by unscrewing the nut holding the cable to the shoe control lever. Free the release spring of the clutch lever and disconnect the cable. Loosen the threaded cap protecting the tie rod attachment. Take out the nut, lock nut, and the bushing. Loosen the five screws and detach the outer cable control support from the cover. The disconnected cover carries the two clutch control levers (inner and outer).

9. Remove the clutch disks and drive the clutch rod into the tooled center hole. Screw a bolt in place to push the spring assembly in far enough so that the disk retaining ring in the bell inner groove is freed and can be driven out. After removing the ring, pull out the fork, and take out the disks, cups, and the springs. **CAUTION:** *The parts are still under compression.*

10. Loosen the lock nut on the clutch drum notches and slide the clutch drum off; if necessary, force a screwdriver between the clutch slotted drum and the bell.

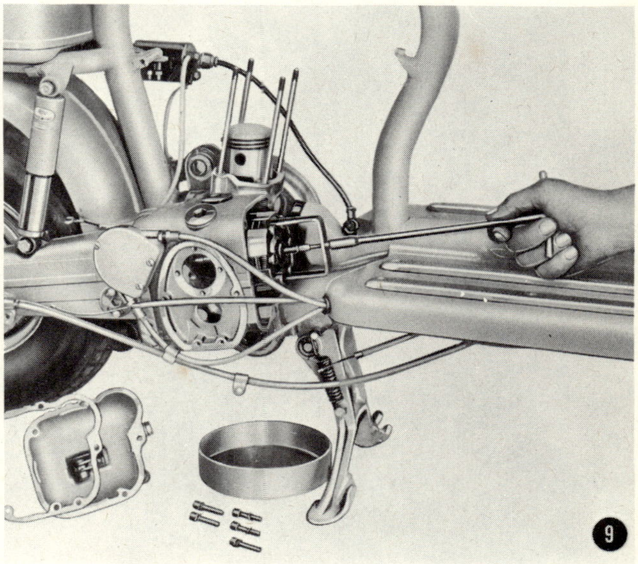

Courtesy Lambretta Division, Innocenti Corporation.

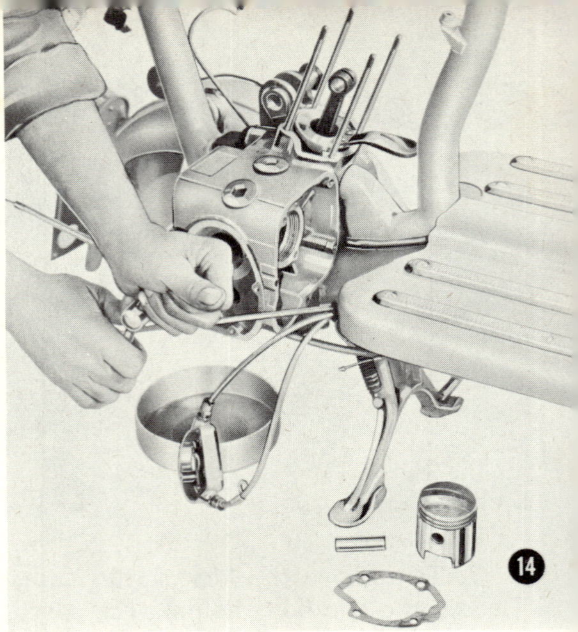

11. Unscrew the bell locking nut by inserting the hub locking wrench through the kick starter bevel gear hole.

12. Apply the bell puller by engaging its six teeth in the grooves of the disk retaining ring. Then operate the puller screw, sliding the bell from the clutch control gear.

13. To remove the piston, first free the piston pin from its lock ring and then pull out the pin.

14. To disassemble the crankshaft, turn it until the connecting rod is at TDC. Insert a wrench in the crankcase hole on the cylinder side to pre-

Courtesy Lambretta Division, Innocenti Corp.

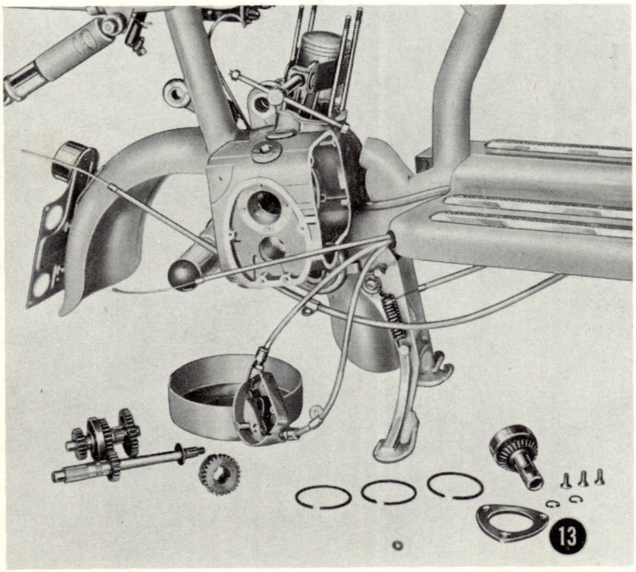

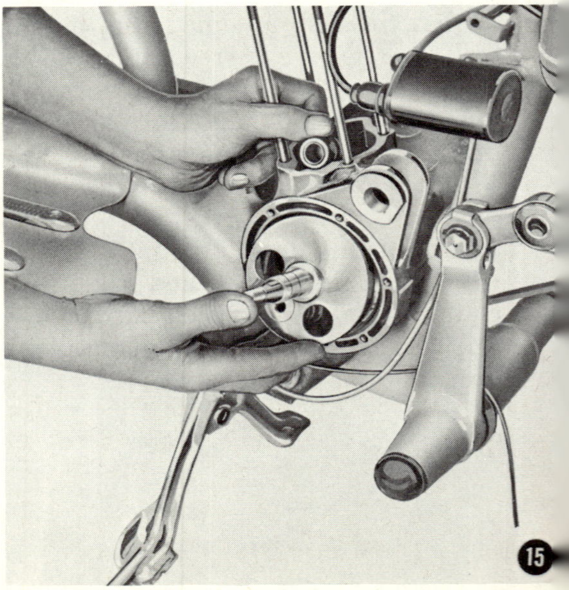

vent rotation of the crankshaft. Then unscrew the lock nut and remove the bevel pinion.

15. Turn the crankshaft until the connecting rod is at BDC. Pull the crankshaft (flywheel magneto side) and push on the connecting rod. Tilt it for easier removal.

16. Clean and inspect all parts. Replace those which show wear. Use new gaskets in assembling. The assembly is the reverse order of disassembly.

A one-cylinder, two-stroke cycle engine is frequently used to power lawnmowers. 1. Clutch control rod; 2. Drive roller; 3. Muffler. Courtesy Lawn-Boy Division, Outboard Marine Corporation.

Parts of the lawnmower and engine assembly. 1. Choke knob and on-off switch; 2. Filler cap; 3. Choke rod; 4. Fuel tank; 5. Fuel filter; 6. Carburetor mixture adjustment knob; 7. Carburetor; 8. Air filter; 9. Reed valve; 10. Crankcase; 11. Height adjustment holes; 12. Blade; 13. Blade stiffener; 14. Muffler; 15. Crankshaft; 16. Exhaust port; 17. Spark plug; 18. Piston; 19. Connecting rod; 20. Cylinder; 21. Armature plate; 22. Flywheel; 23. Governor; 24. Starter. Courtesy Lawn-Boy Division, Outboard Marine Corporation.

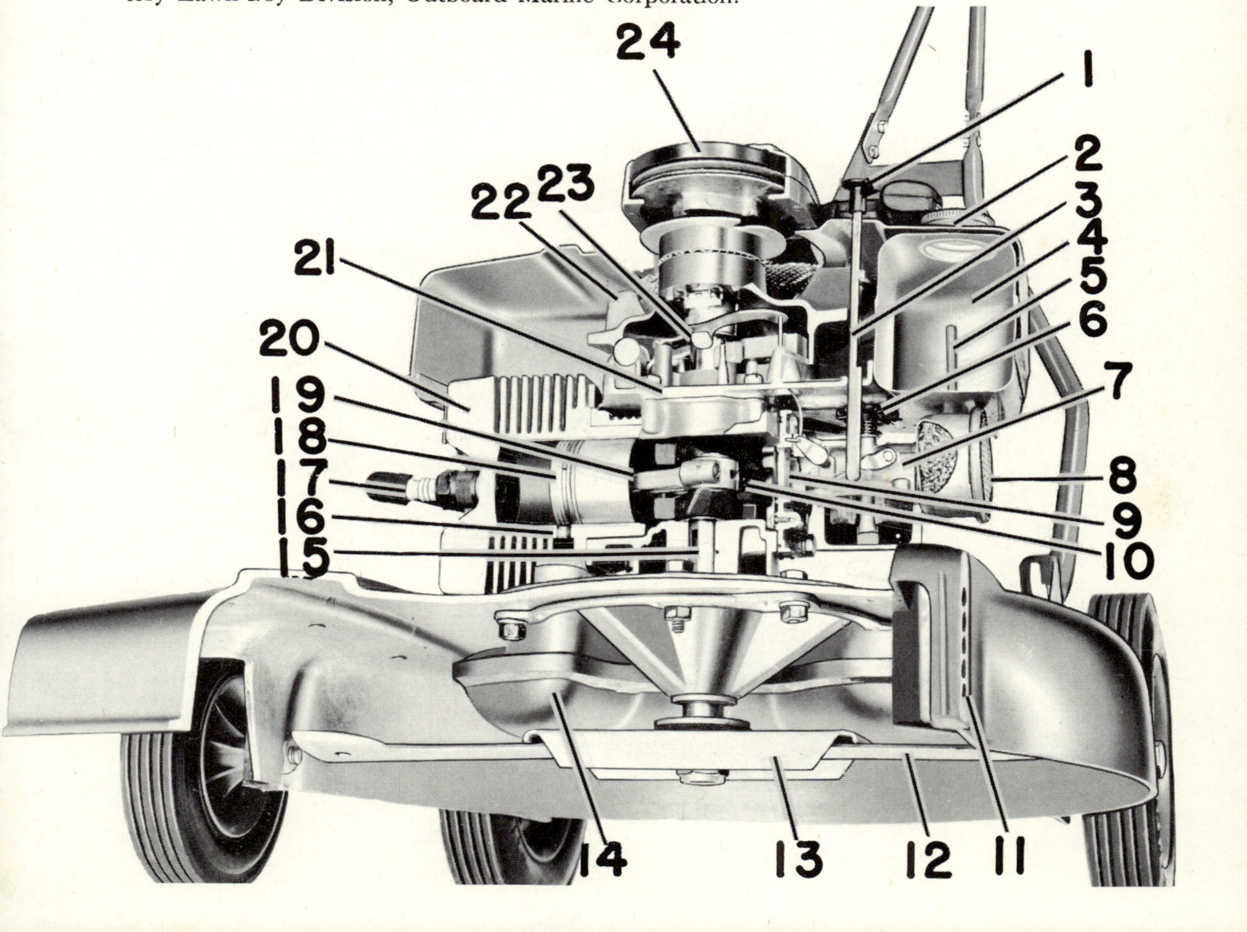

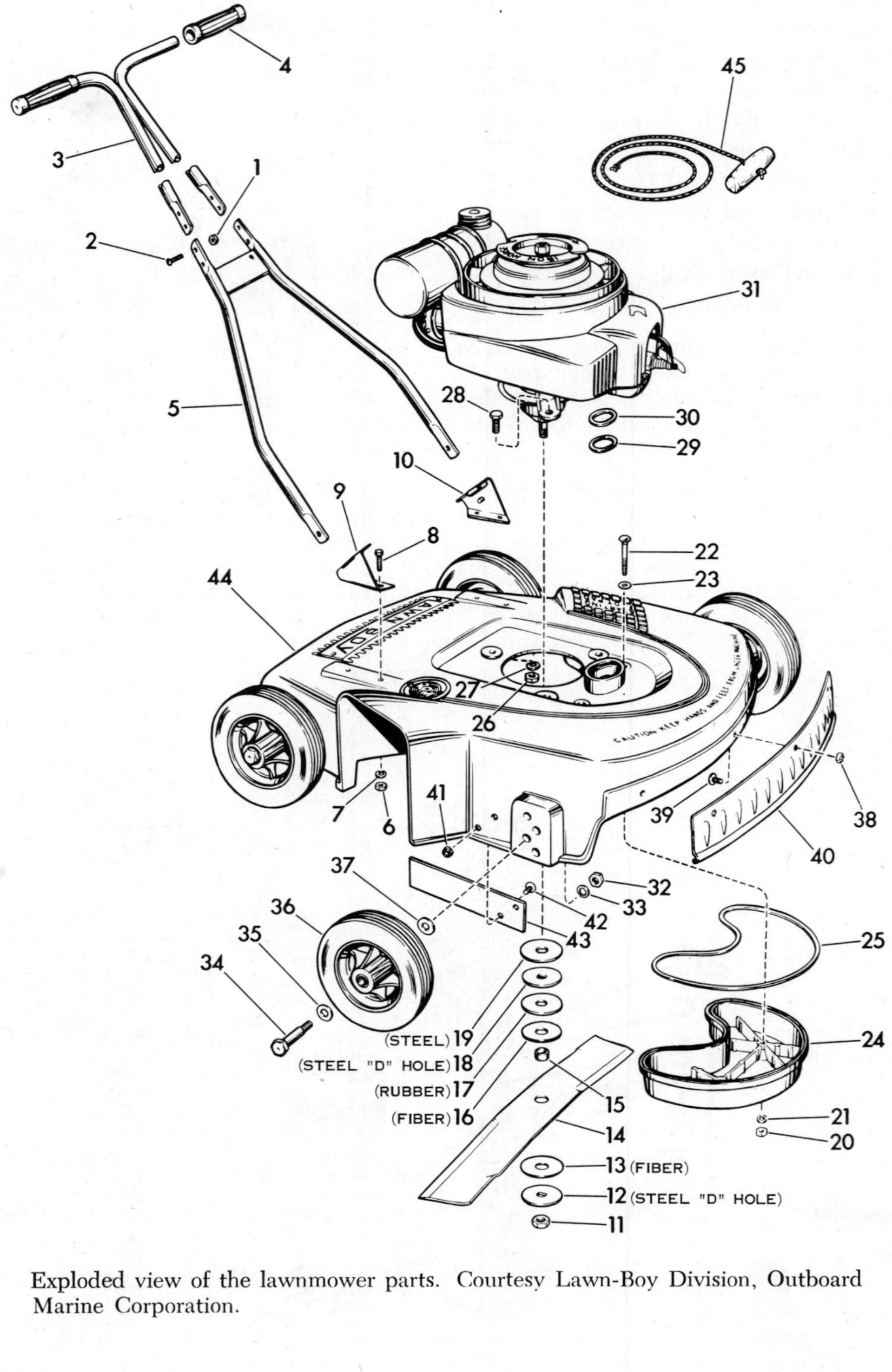

Exploded view of the lawnmower parts. Courtesy Lawn-Boy Division, Outboard Marine Corporation.

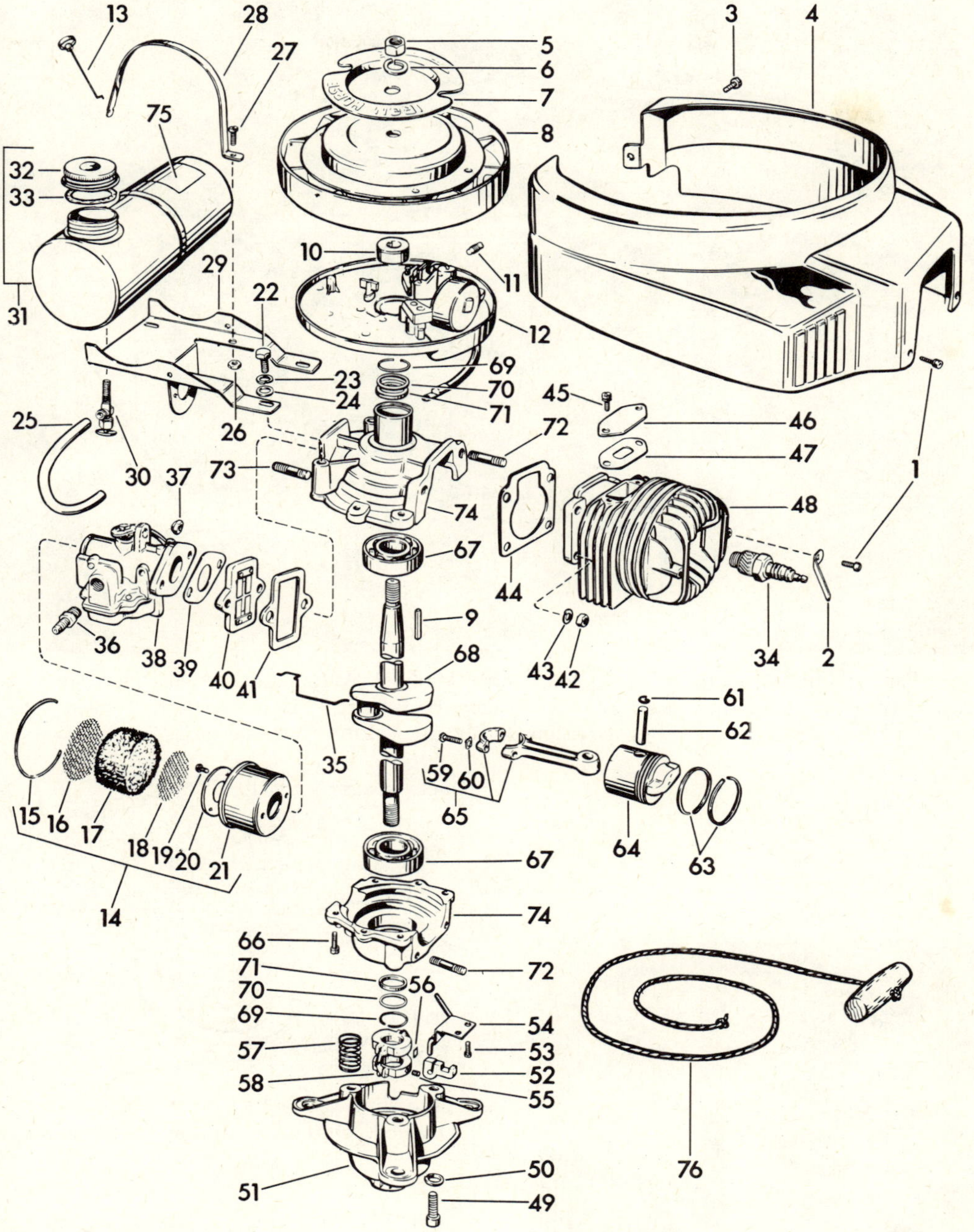

Exploded view of the engine parts. Courtesy Lawn-Boy Division, Outboard Marine Corporation.

Parts of the power drive. 1. Speed control knob; 2. Automatic lock; 3. Variable drive pulley; 4. Idler gear; 5. Driver roller; 6. V-belt; 7. Drive gear; 8. Driven pulley. Courtesy Lawn-Boy Division, Outboard Marine Corporation.

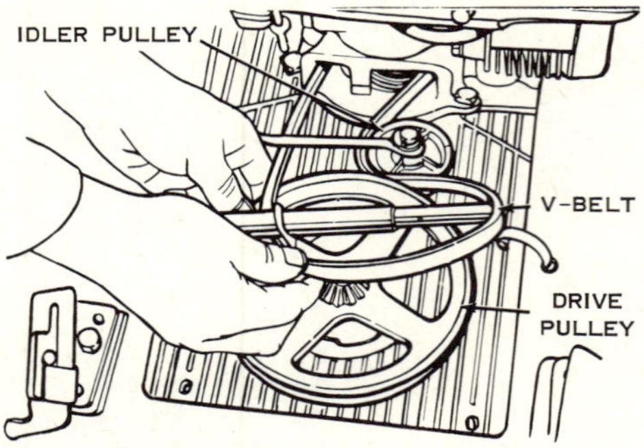

Threading the belt. Courtesy Lawn-Boy Division, Outboard Marine Corporation.

An off-center blade will vibrate. Check for trueness by marking a straight edge where the blade ends. Then turn the blade one half turn, and the marks should line up. Courtesy Lawn Boy Division, Outboard Marine Corporation.

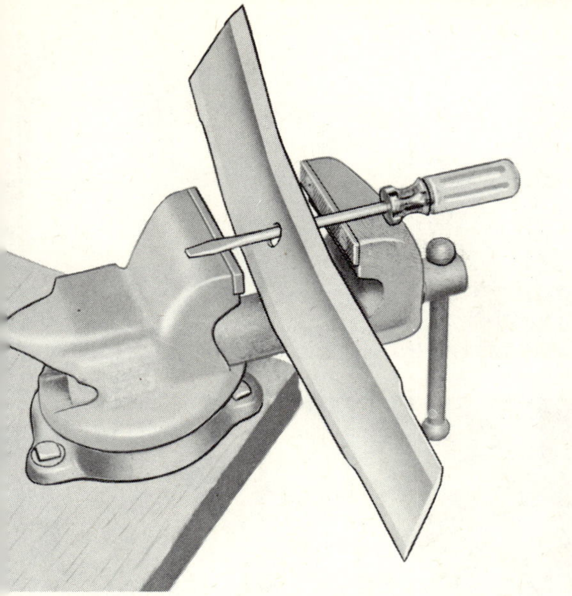

Testing the blade balance. Grind material off the heavy side. Courtesy Lawn-Boy Division, Outboard Marine Corporation.

DISASSEMBLING A TWO-CYCLE, TWO-CYLINDER ENGINE

(Each of the following steps is illustrated. The number in the lower right-hand corner of the illustration agrees with the numbered steps below.)

1. Ready to go to work. Note the orderly array of tools, the clamping device to hold the engine securely, and the general cleanliness of the work area.
2. Remove the shroud.
3. Remove the starter housing retaining screw.
4. Remove the starter housing assembly.

Courtesy Johnson Motors Division, Outboard Marine Corporation.

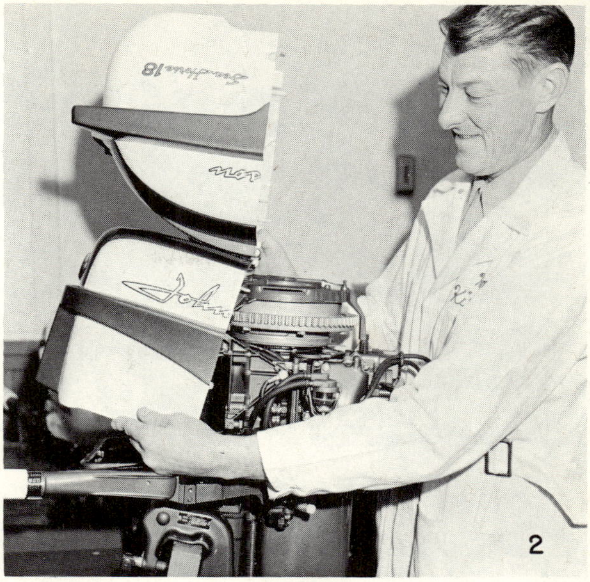

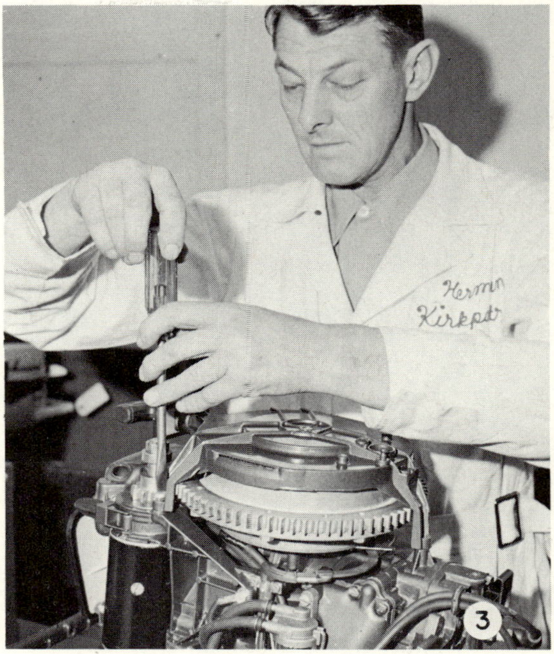

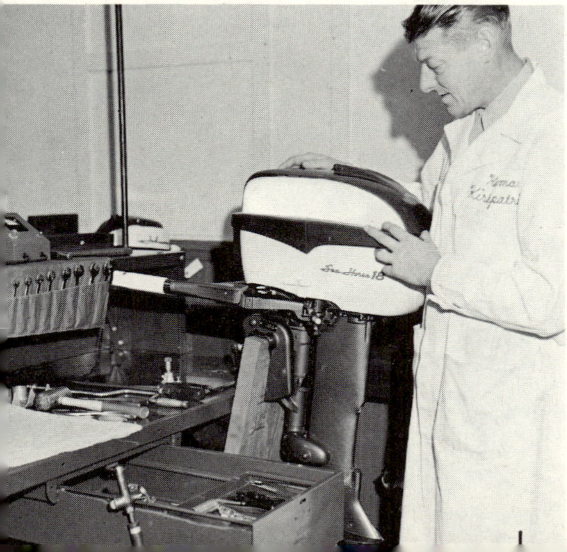

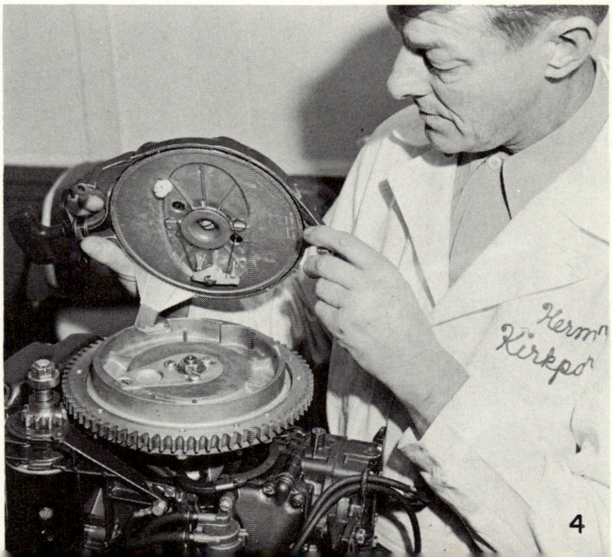

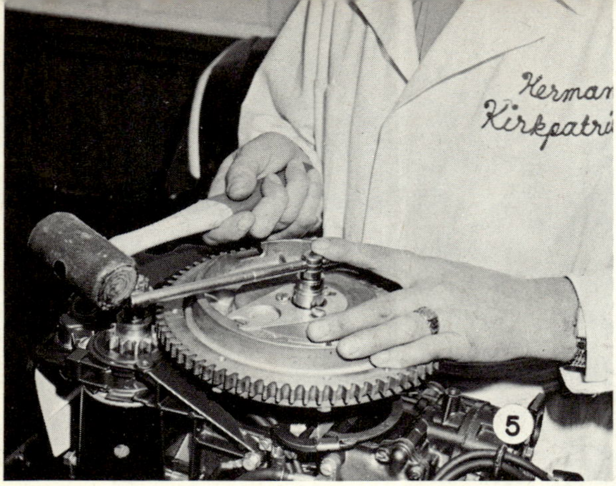

5. Remove the flywheel nut.
6. Remove the flywheel cover.
7. Use a puller to remove the flywheel.

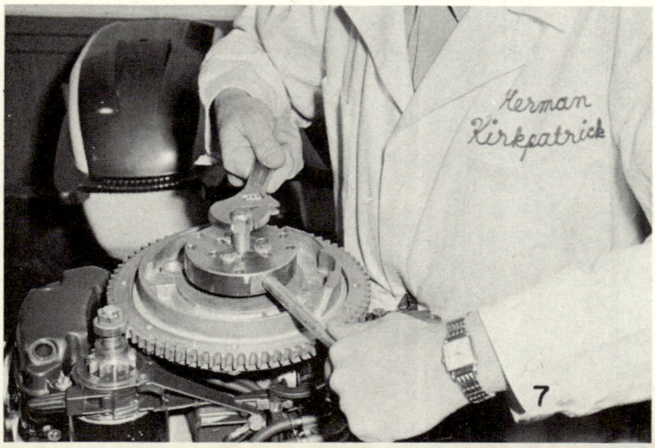

8. Lift the flywheel straight up.
9. Remove the spring clip from the link and stop assembly.
10. Remove the magneto plate assembly.

Courtesy Johnson Motors Division, Outboard Marine Corporation.

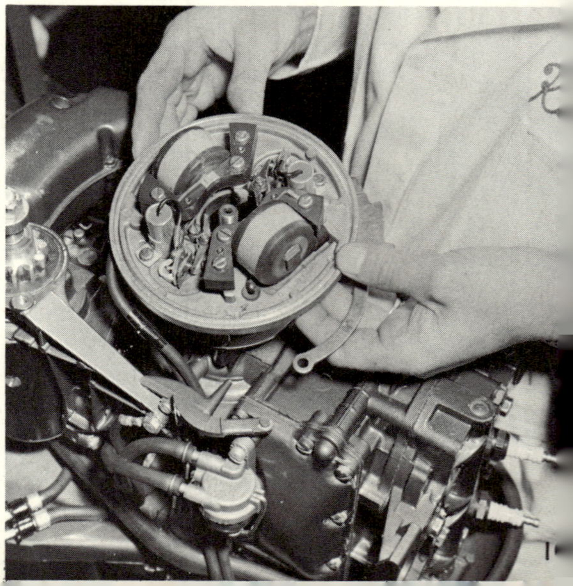

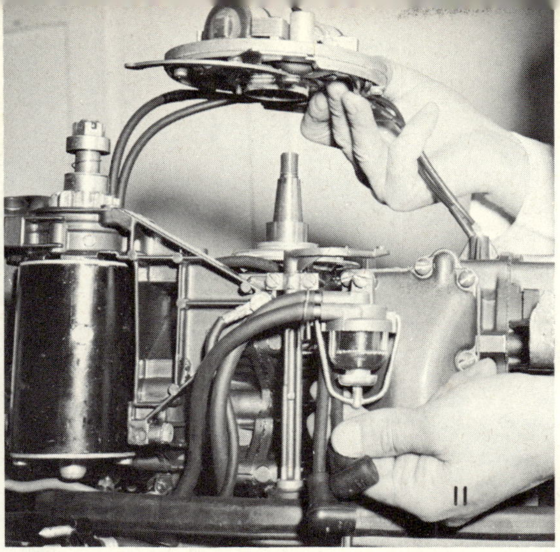

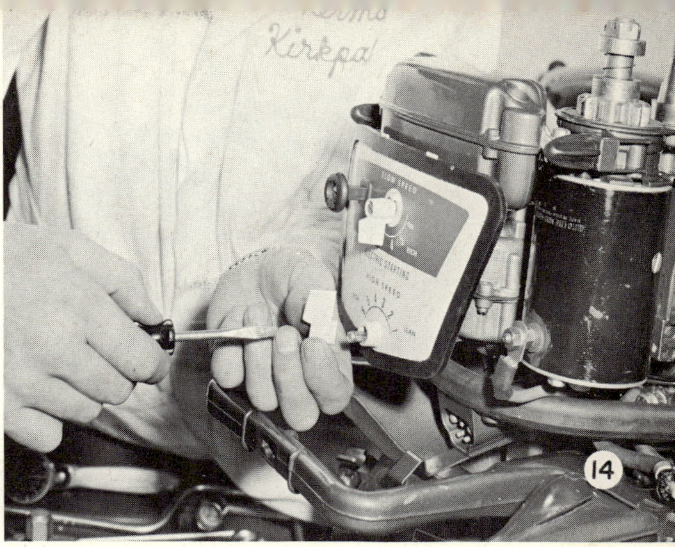

11. Remove the spark plug wires from the bracket.

12. Remove the screws holding the support plate and retaining ring.

13. Remove the support plate and retaining ring.

14. To remove the control panel, take off the high-speed adjusting lever, the low-speed lever, and the choke knob.

15. Remove the air silencer.

16. Remove the spark plugs.

Courtesy Johnson Motors Division, Outboard Marine Corporation.

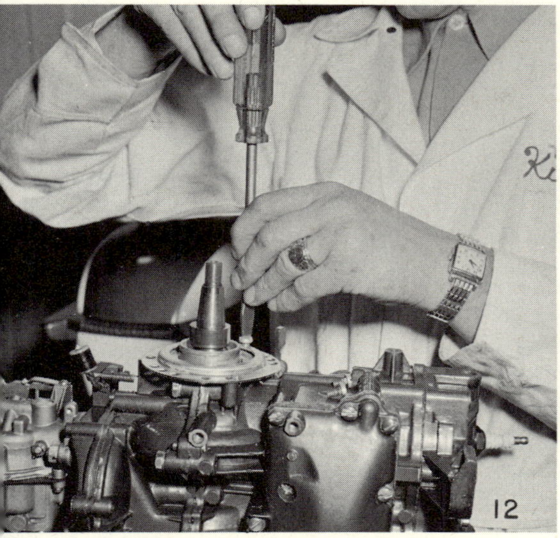

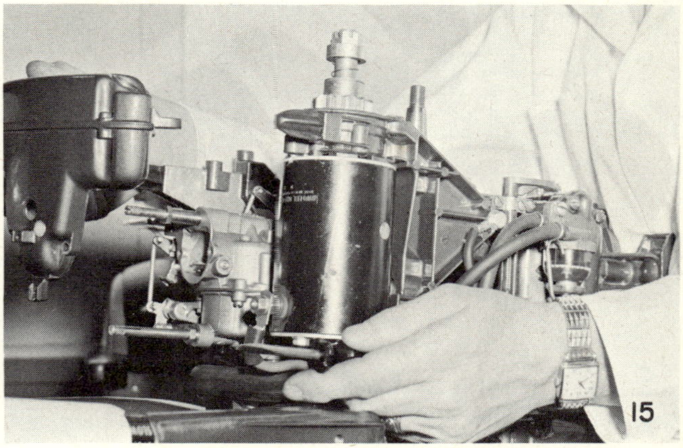

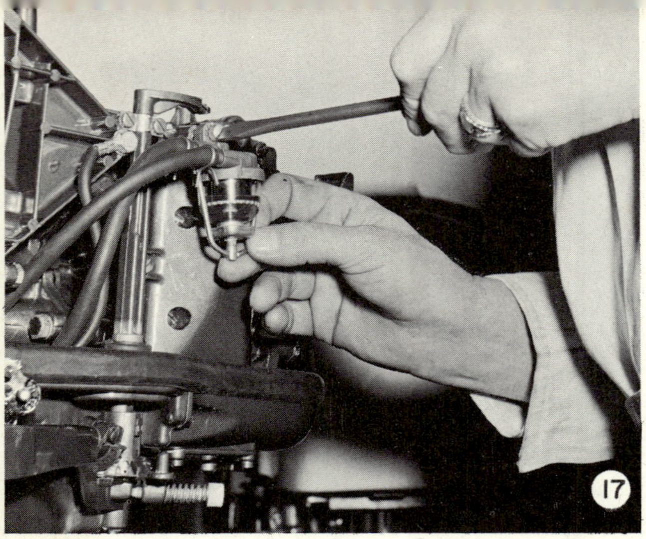

17. Remove the fuel filter.

18. Remove the electric starting motor assembly.

19. Remove the starter and throttle lever as an assembly.

20. Remove the carburetor.

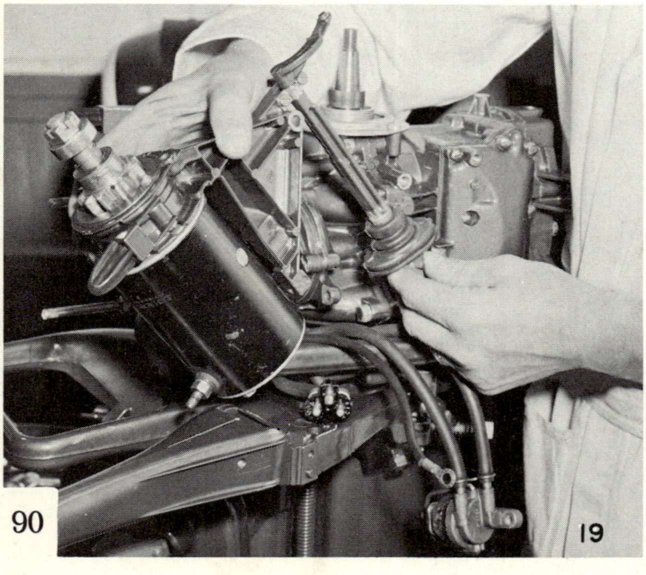

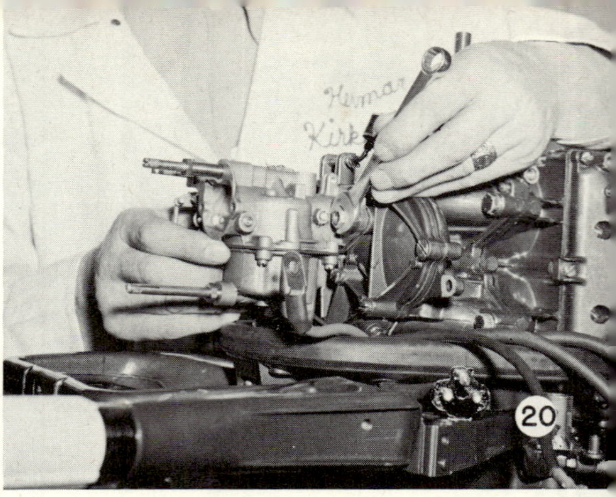

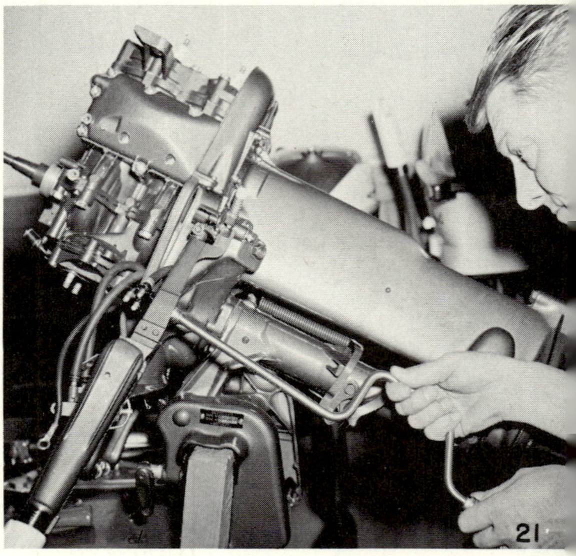

21. Remove the male fuel connector assembly which is held in place by a screw below the casting.

22. Remove the lower front cover as an assembly, keeping the cable assembly connector, filter bowl, male fuel connector, and lines attached.

Courtesy Johnson Motors Division, Outboard Marine Corporation.

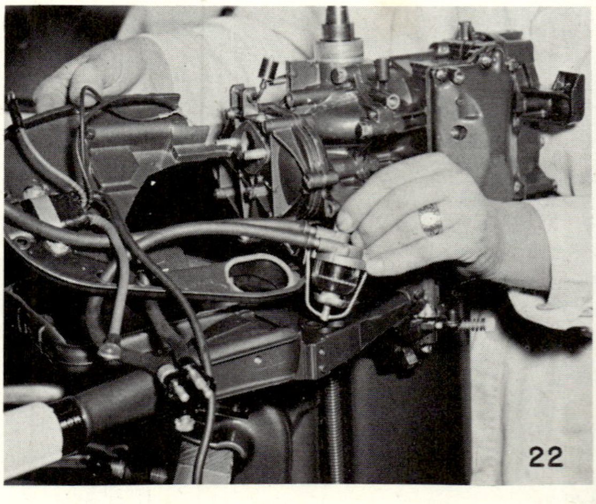

23

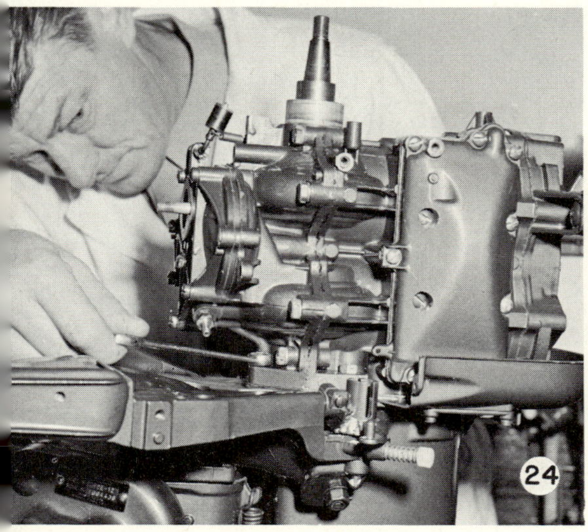

24

23. Remove the shifter lock and spring.

24. To separate the power head from the lower unit, remove the bolts from the top and the screws from below.

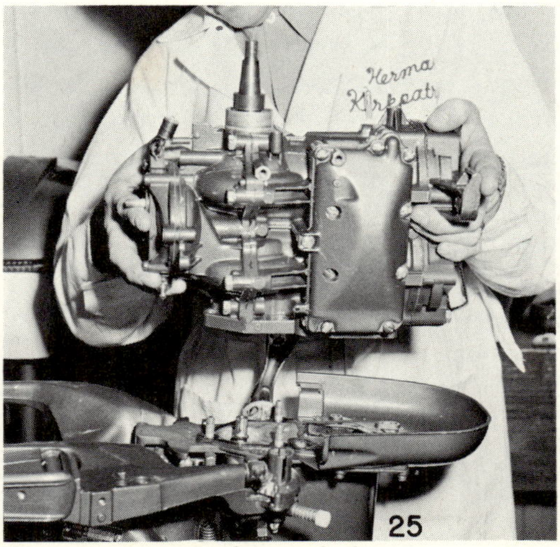

25

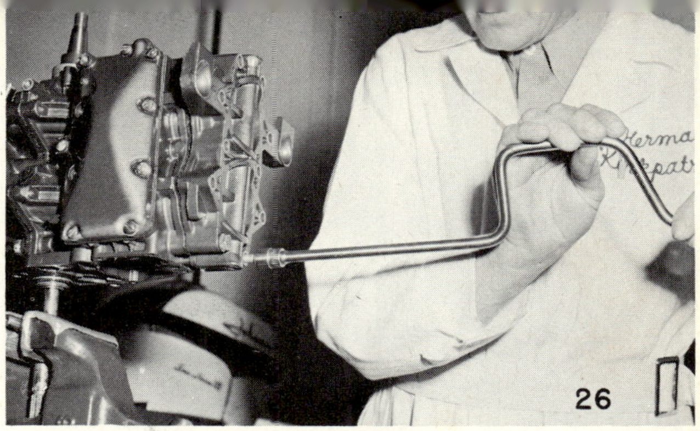

26

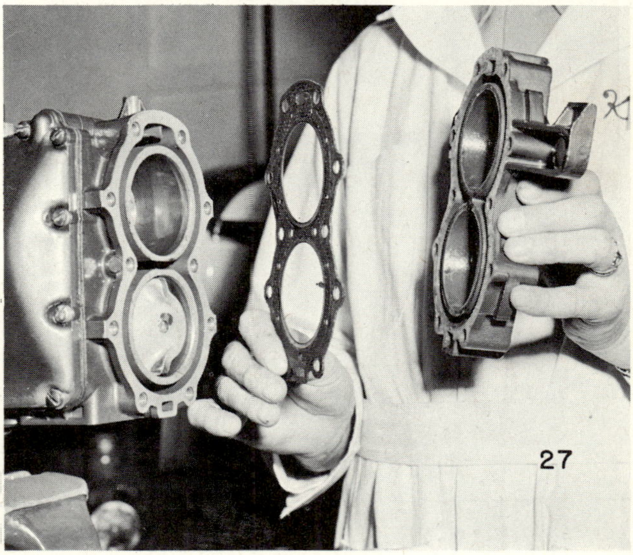

27

25. Separate the power head from the lower unit.

26. Remove the cylinder head bolts.

27. Remove the cylinder head and gasket from the power head.

28. Remove the intake manifold cover screws.

Courtesy Johnson Motors Division, Outboard Marine Corporation.

28

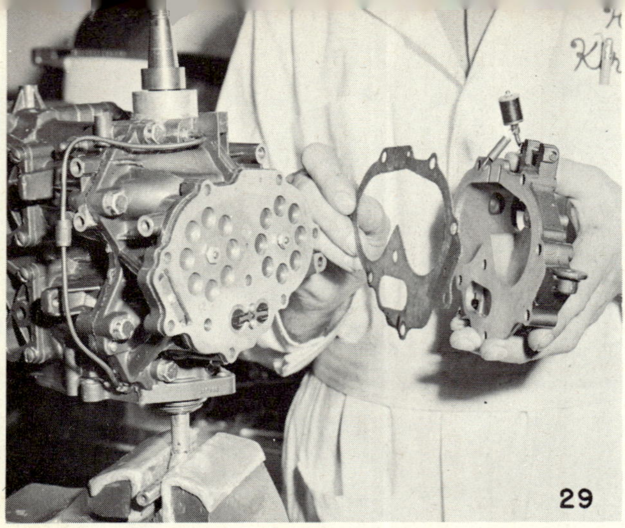

29

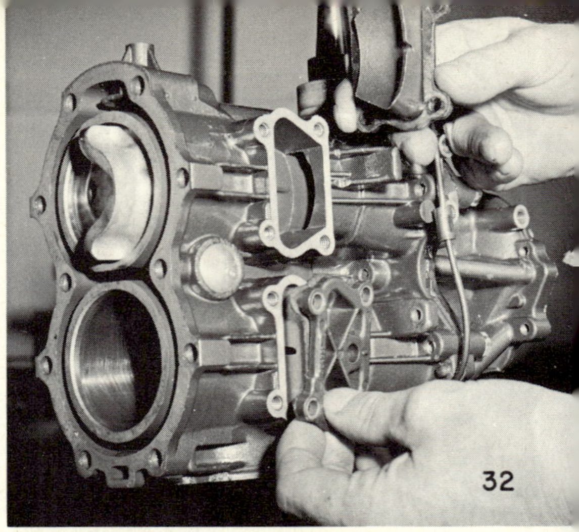

32

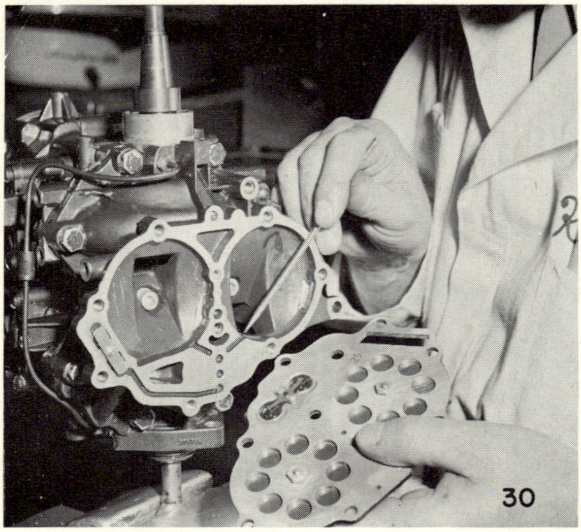

30

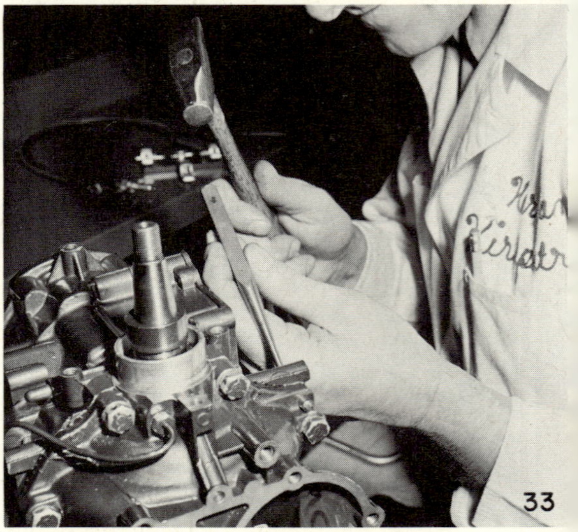

33

29. Remove the intake manifold cover to expose the leaf plate assembly.

30. Remove the leaf plate. The pointer shows the passage which releases crankcase pressure to the remote pressurized fuel tank.

31. Remove the outer exhaust cover and gasket (right) and the inner exhaust cover and gasket.

32. Remove the intake by-pass covers and gaskets (both top and bottom).

33. Drive out the tapered line-up pins. (Drive toward the carburetor.) **Note:** *These pins line up the crankcase and cylinder.* They must be removed before the retaining screws are taken out, and they must be installed first to make sure that the halves are properly mated during assembly.

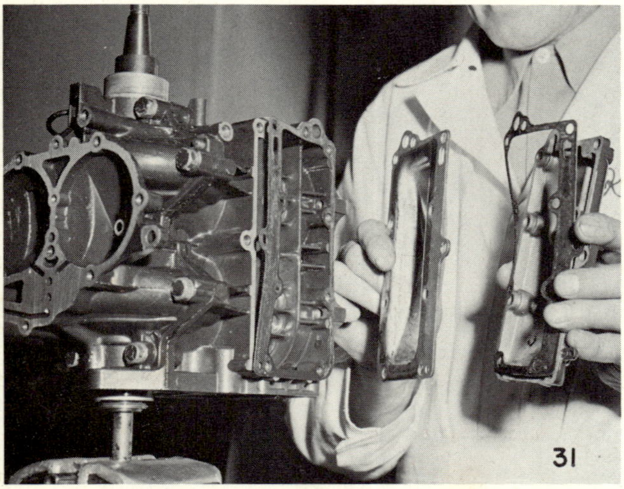

31

Courtesy Johnson Motors Division, Outboard Marine Corporation.

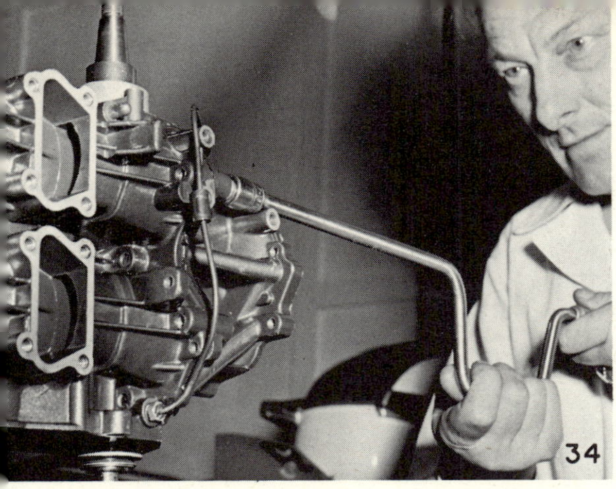

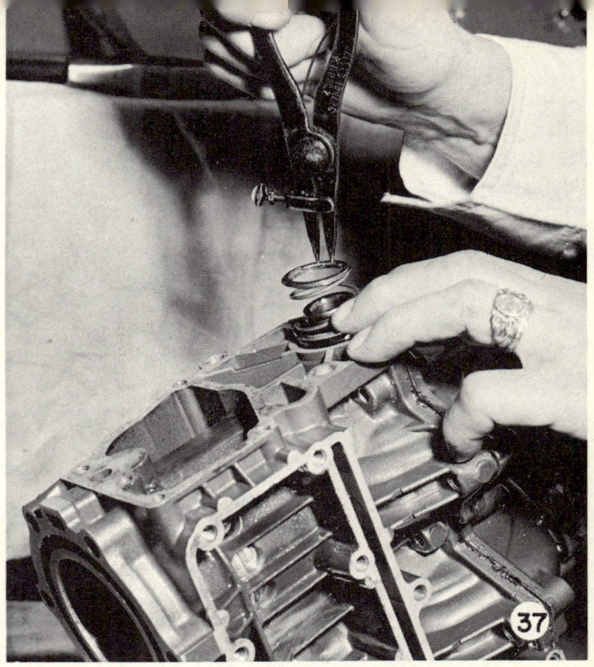

34. Remove the crankcase-to-cylinder bolts.

35. Loosen the center Allen screws which hold the crankcase to the cylinder block.

36. Remove the crankcase.

37. Remove the retaining rings, washers, and springs from the bottom of the crankshaft.

38. Remove the connecting rod cap screws.

Courtesy Johnson Motors Division, Outboard Marine Corporation.

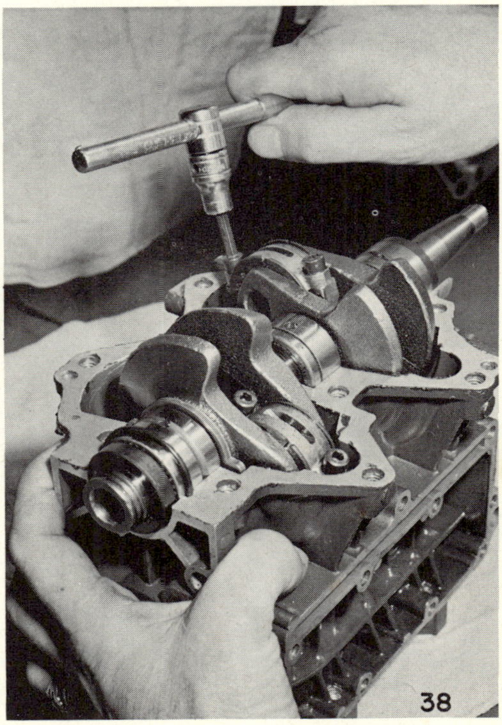

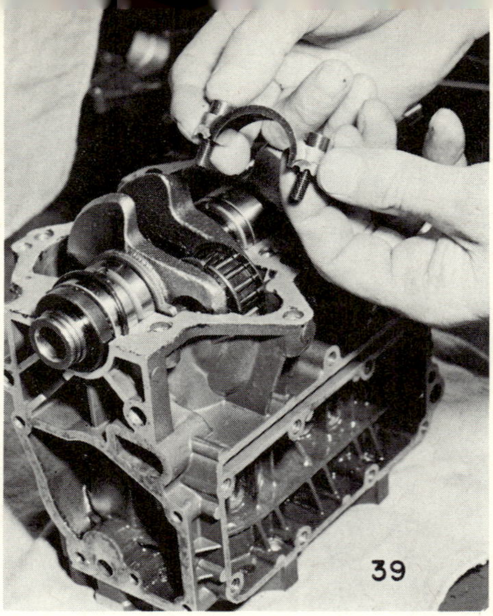

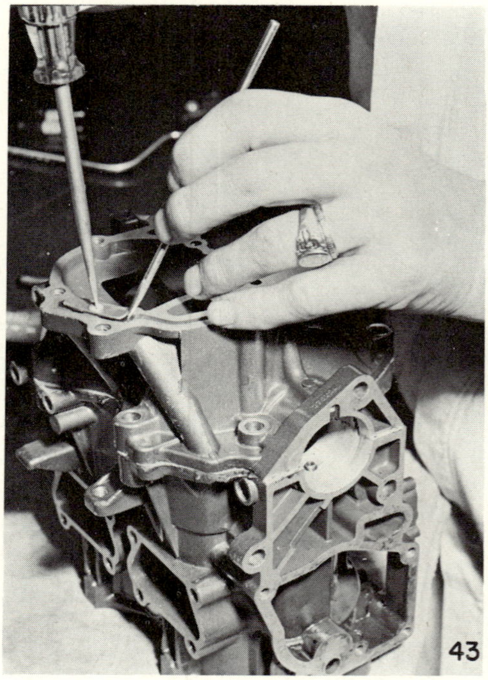

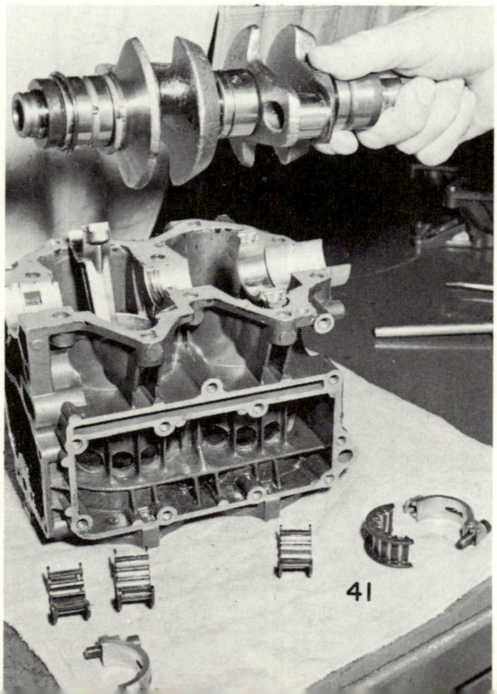

39. Lift the cap to expose the cage and needle bearings.

40. Remove the crankshaft.

41. This illustration shows the proper way to separate the units of the matched cages and connecting rod caps so that they can be properly reassembled. Note the clean rag to minimize dirt contamination.

42. Remove the piston, rod, and ring assembly from the cylinder.

43. Check to see that the leaf valve is seating properly.

Courtesy Johnson Motors Division, Outboard Marine Corporation.

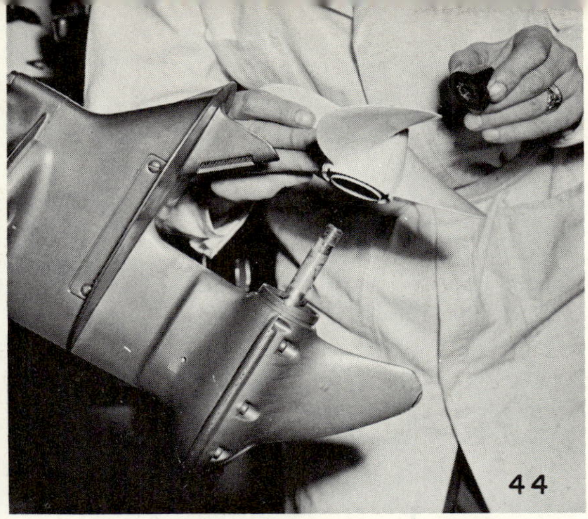

44

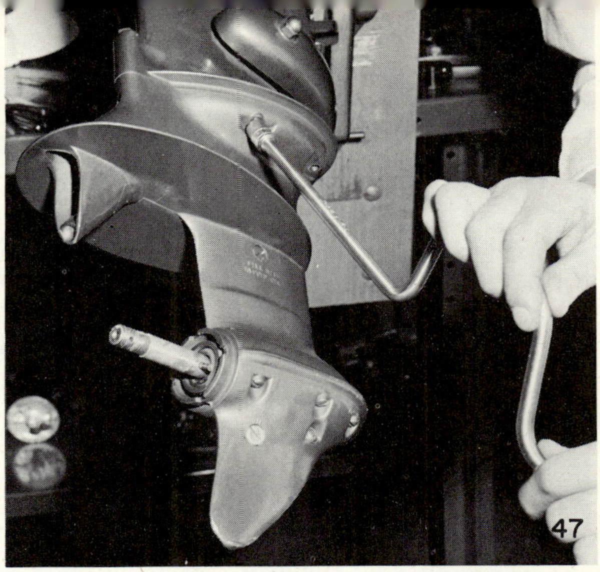

47

45

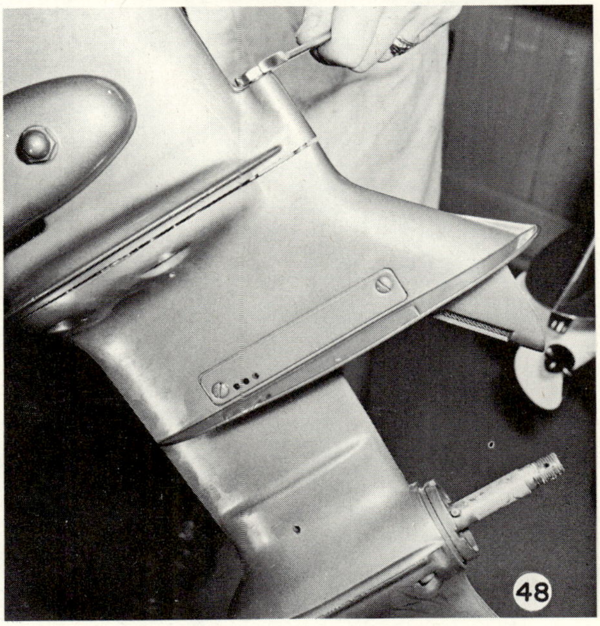

48

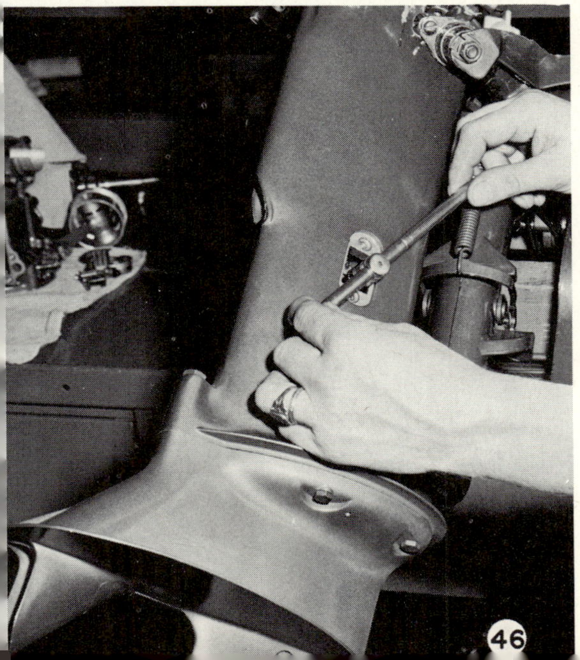

46

Disassembling the Lower Unit

44. Remove the retainer and the propeller.

45. Remove the top and bottom oil plugs to drain the oil.

46. Remove the cover plate and take out the bottom shift connector screw.

47. Remove the four screws holding the gear case to the exhaust housing.

48. Remove the bolt holding the exhaust housing to the gear case.

Courtesy Johnson Motors Division, Outboard Marine Corporation.

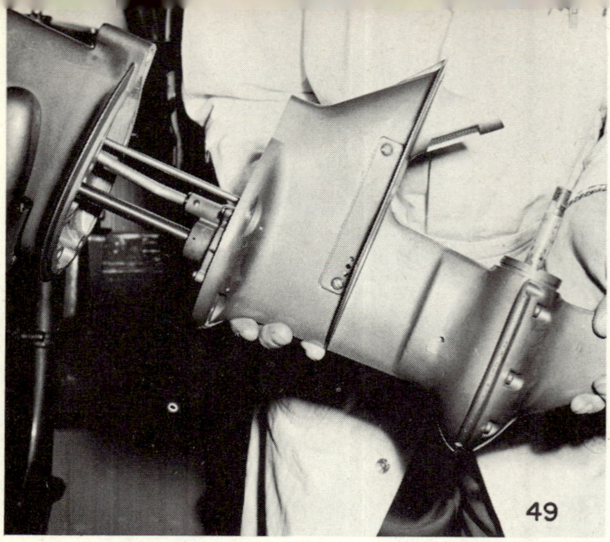

49

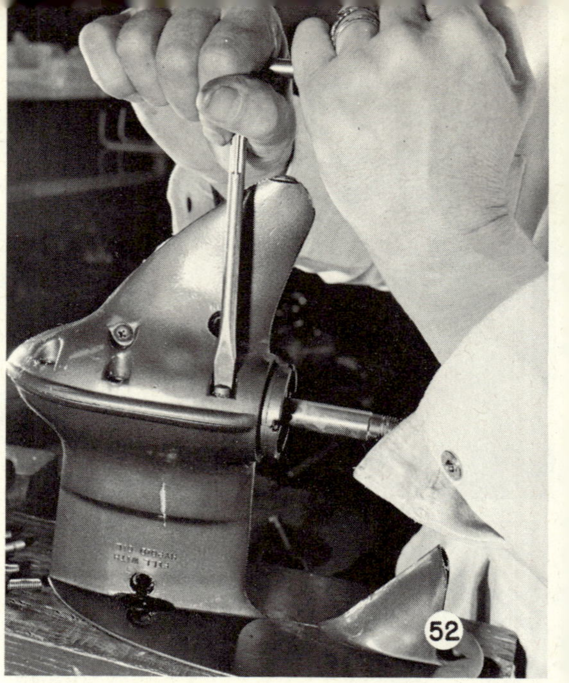

52

50

49. Pull the gear case, water pump assembly, and the water tube from the exhaust housing.

50. Remove the pump housing screws.

51. Lift the pump housing, impeller, plate, and drive shaft from the gear case.

52. Remove the screws holding the skeg to the gear case.

53. Remove the pivot screw with a Phillip's screwdriver.

Courtesy Johnson Motors Division, Outboard Marine Corporation.

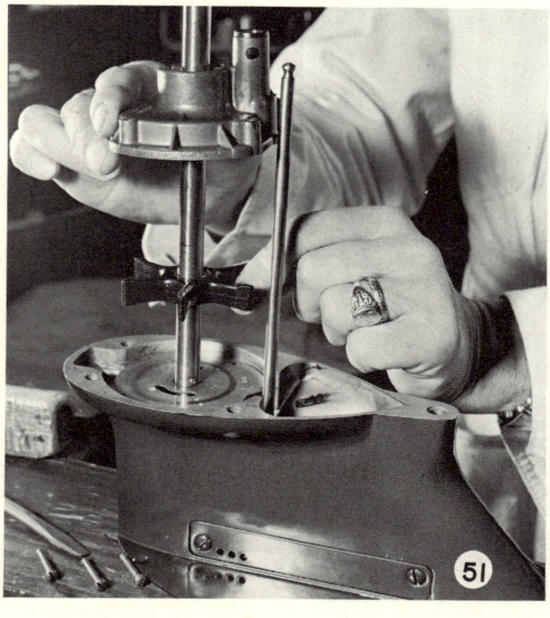

51

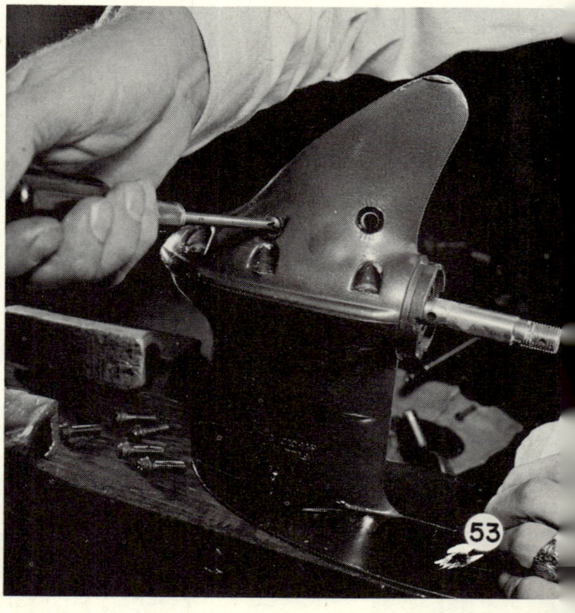

53

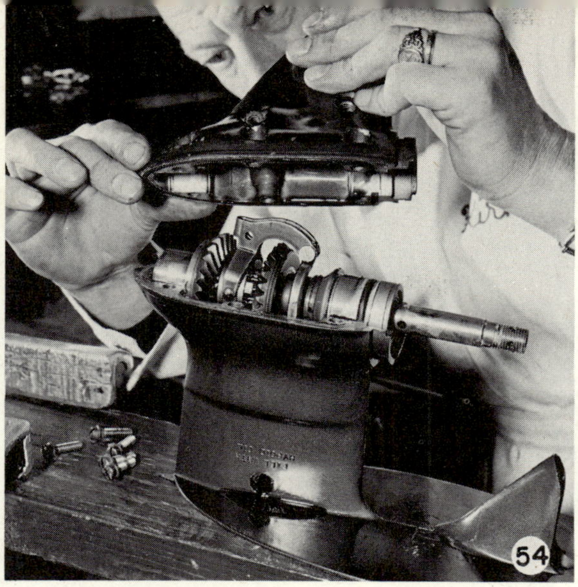

54

55

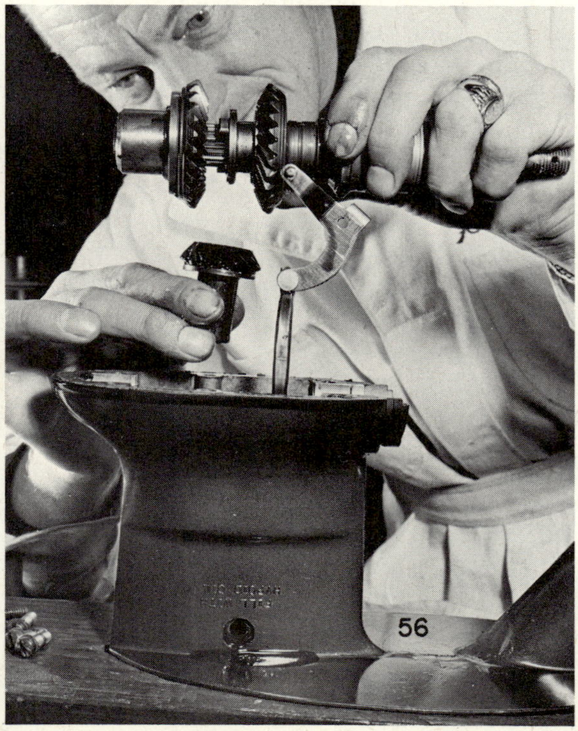

56

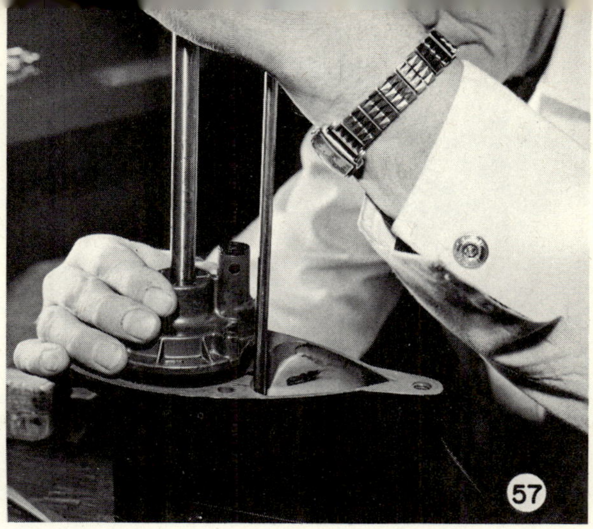

57

54. Remove the skeg from the gear case housing.

55. Remove the propeller shaft, forward, and reverse gear assembly.

56. Remove the pinion gear.

Clean all parts and inspect for wear.

Assembling

(In general, the assembly is the reverse of disassembly. However, there are some special details which are shown in the following illustrations. Not all assembly steps are shown.)

57. Install the pump housing by pushing down and turning the shaft.

58. Fill the gear case through the bottom drain plug with hy-point oil. Continue filling until the oil flows from the top plug. Hold the nozzle firmly in the bottom plug until the top plug is installed. Then remove the nozzle and install the bottom plug.

Courtesy Johnson Motors Division, Outboard Marine Corporation.

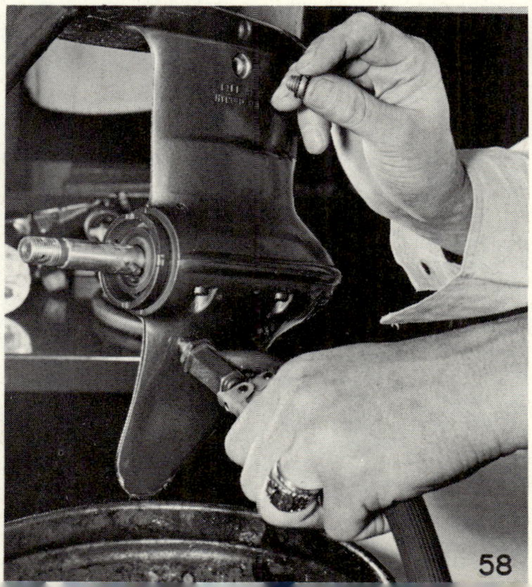

58

59

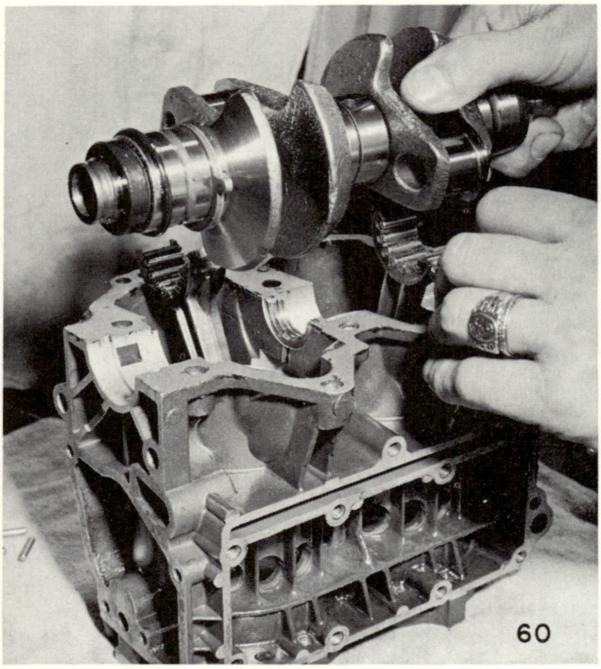

60

61

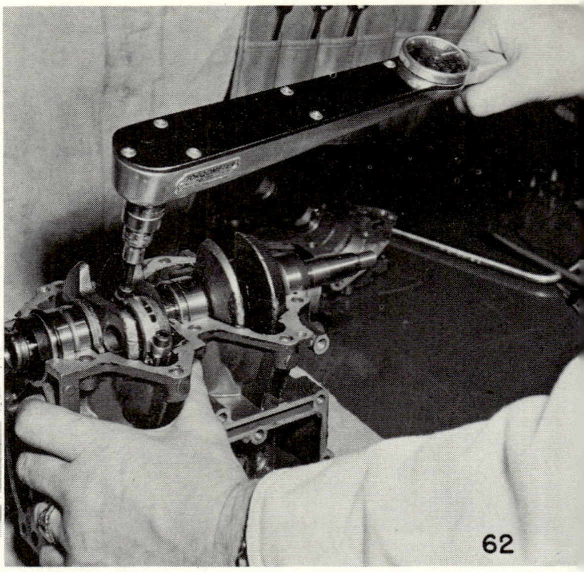

62

63

59. Oil the cylinder bores and ring grooves. Install the piston, rod, and ring assembly. Be sure that the rings are in their correct position before pressing the piston into the cylinder.

60. Use grease to lubricate the connecting rods. Reassemble the cages and needle bearings. Lay the crankshaft into the connecting rods.

61. Assemble the connecting rod cap, cage, and needle bearings. Be sure that the cap and cage are matched.

62. Torque the connecting rod screws to 15 ft. lbs.

63. Install the flywheel retaining nut and torque it to 45 ft. lbs.

Courtesy Johnson Motors Division, Outboard Marine Corporation.

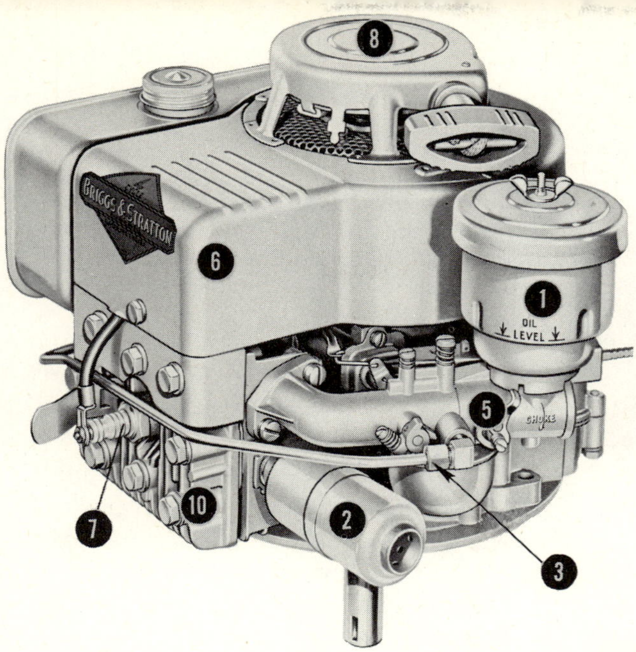

A horizontal, four-stroke-cycle, one-cylinder engine. The numbers correspond to the numbered instructional steps of the text. Courtesy Briggs and Stratton Corporation.

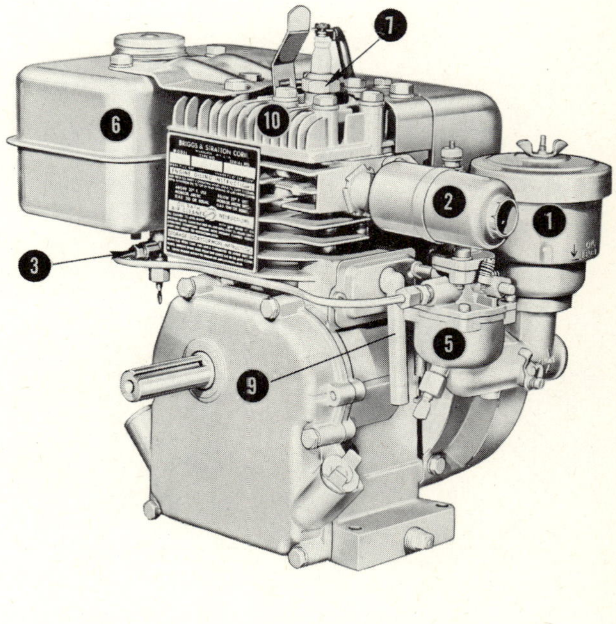

A vertical, four-stroke-cycle, one-cylinder engine. The numbers correspond to the numbered instructional steps of the text. Courtesy Briggs and Stratton Corporation.

DISASSEMBLING A FOUR-CYCLE, ONE-CYLINDER ENGINE

1. Remove the air cleaner.
2. Remove the muffler.
3. Remove the fuel lines and connections.
4. Remove the fuel filter.
5. Remove the carburetor linkage, and then the carburetor.
6. Remove the blower housing and fuel tank.
7. Remove the spark plug.
8. Remove the rope starter pulley.
9. Remove the breather pipe.
10. Remove the cylinder head.
11. Remove the flywheel.
12. Remove the magneto.
13. Remove the stator plate assembly.
14. Remove the bearing plate.
15. Remove the oil pump.
16. Remove the governor shaft and governor.
17. Remove the connecting rod and piston assembly.

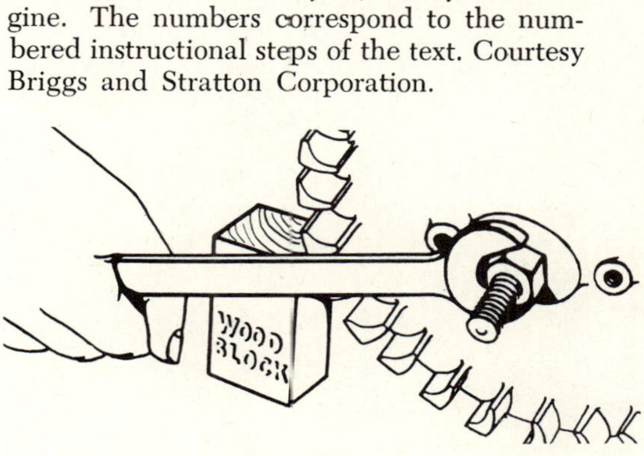

Removing the crankshaft nut can be done easier if the flywheel can be kept from rotating. Courtesy Briggs and Stratton Corporation.

Tap the nut lightly while keeping pressure against the flywheel, in order to loosen it from the taper. Courtesy Lawn-Boy Division, Outboard Marine Corporation.

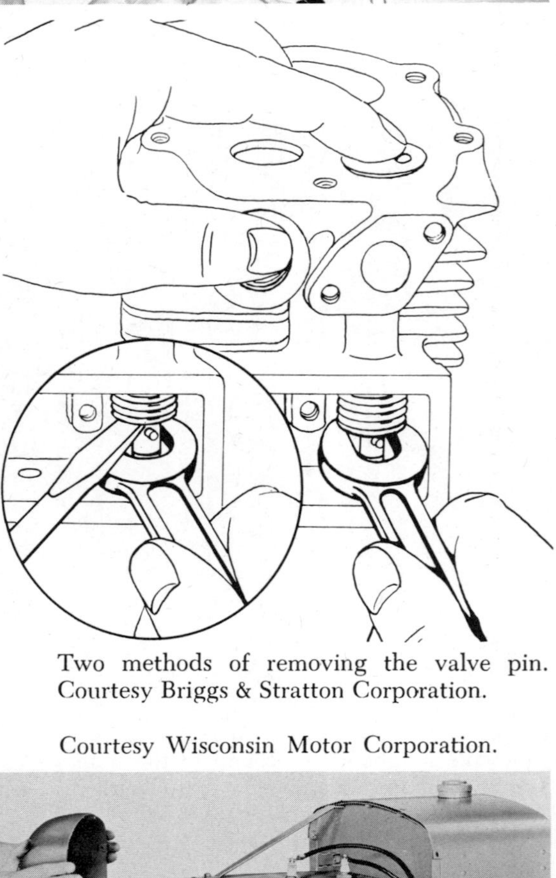

Two methods of removing the valve pin. Courtesy Briggs & Stratton Corporation.

Courtesy Wisconsin Motor Corporation.

18. Remove the valve locks, valves, and valve springs.

19. Remove the camshaft gear.

20. Remove the crankshaft assembly.

21. Clean all parts and inspect for wear. Replace those parts which show wear. Use new gaskets to seal joints. Dress down mating parts to be sure that they are flat.

22. The assembly is the reverse of disassembly.

DISASSEMBLING A FOUR-CYCLE, TWO-CYLINDER ENGINE

(Each of the following steps is illustrated. The number in the lower right-hand corner of the illustration agrees with the numbered steps below:)

1. Remove the muffler and disconnect the governor control, choke, choke control, ignition switch, fuel lines and, if an electric starter and generator are used, these should also be removed. The top or canopy is removed by taking out the screws holding it to the end panels. Remove the air cleaner and bracket, the flywheel screen, and the four round-headed screws which support the panel to the side rails. The front housing panel can now be lifted off.

2. The fuel tank assembly can be removed by taking out the six screws which hold it to the rear panel. The rear housing panel can then be removed by taking out the screws holding it to the cylinder block, engine base, and side rails.

3. Loosen the flywheel nut with a $1\frac{3}{8}$ inch open-end wrench.

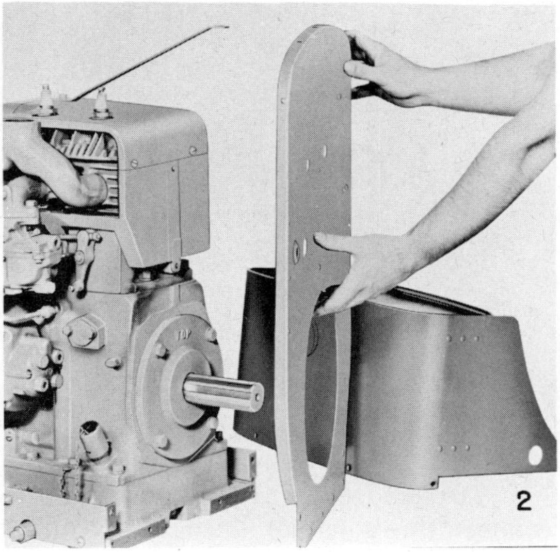

2

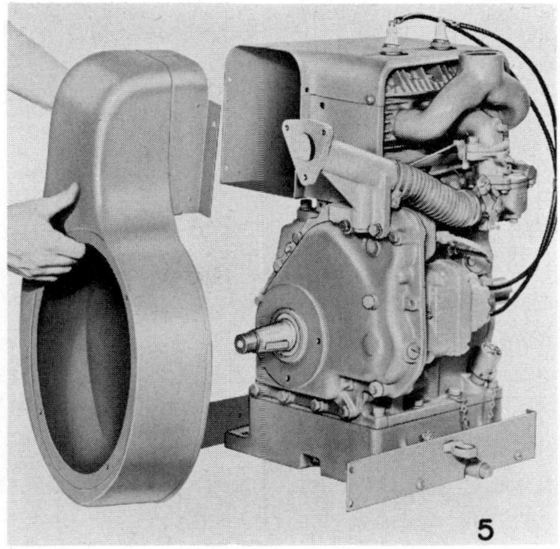

5

3

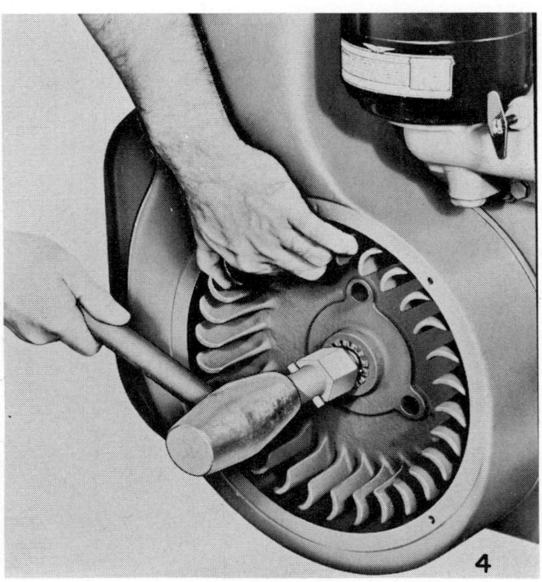

4

6

4. Unscrew the flywheel nut about two turns but do not remove it. The flywheel is mounted on a taper. Take a firm hold on the flywheel fins, pull outward and, at the same time, strike the end of the flywheel nut with a babbitt hammer The flywheel will slide off the taper.

5. The three capscrews which mount the flywheel shroud to the gear cover and the six screws to the cylinder shrouding must be removed to disassemble the flywheel shroud from the engine.

6. To remove the carburetor and

Courtesy Wisconsin Motor Corporation.

manifold, remove the cotter pin from the governor control rod, and pull the rod from the control lever. Take out the two capscrews which hold the air cleaner connection bracket to the gear cover and remove the two nuts and square washers from the manifold studs. The complete carburetor, manifold, and air cleaner connection bracket can be removed as a unit.

7

8

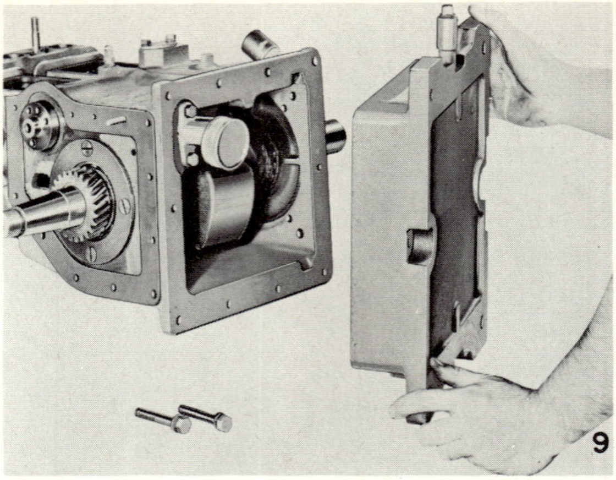

9

7. To remove the gear cover, disconnect the governor linkage and oil line. Remove the governor, magneto, and the gear cover mounting screws. Screw a 5/16 inch SAE capscrew into one of the flywheel shroud mounting holes. Then, by lightly tapping the gear cover through the magneto mounting hole, the gear cover will come off without damage to the dowel pins. Pull out the camshaft thrust plunger and spring to prevent losing them. Remove the cylinder head cover, heat deflector, and side shroud. Remove the cylinder head screws and the head.

8. To remove the camshaft gear, remove the three capscrews which hold the gear to the end of the camshaft, and then pry it off with a screwdriver. Note that the mounting holes in the camshaft gear are staggered in such a manner that the gear can be assembled to the shaft in only one way. The gear cover spacer plate can now be removed.

9. To remove the engine base and oil pump, drain the oil. Take out the eight capscrews which mount the engine base to the crankcase. Then turn the engine on its side and take out the two capscrews that hold the base to the case. Take out the three capscrews which hold the oil pump to the crankcase.

Courtesy Wisconsin Motor Corporation.

10

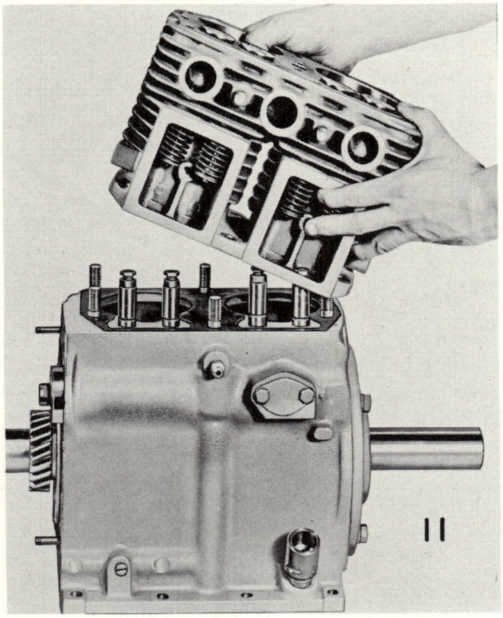
11

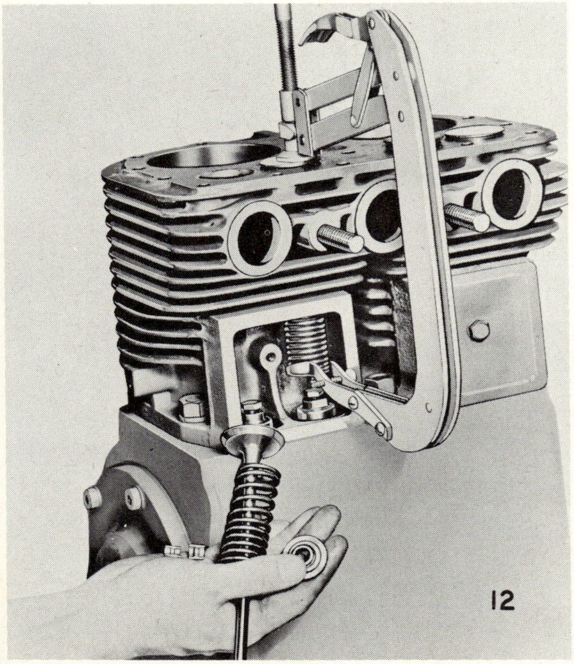
12

10. To remove the pistons and connecting rods, remove the "Palnuts" and hexagon nuts. Then by tapping the ends of the bolts lightly, being careful not to mar the threads, the connecting rod cap can be freed from the bolts. The rod with the piston can now be pushed up through the cylinder. Be careful not to score the crankshaft journals by allowing the rod bolts to strike or scrape across them. Replace the caps on the rods immediately so that they are in the correct position for reassembly making sure that the shims are in place. A number is stamped on the side of the rod and cap for matching purposes.

11. The cylinder block assembly can be removed by taking off the six nuts which hold the block to the crankcase studs.

12. To remove the valves, remove the valve tappet inspection plate, compress the valve springs with a valve lifter, and remove the valve spring retaining locks. Remove the seats, springs, and valves. Clean these parts and the valve ports and guides of all carbon and gum deposits. Tag each valve so that it can be replaced in the same guide during assembly.

13. To remove the crankshaft, take out the four capscrews holding the main bearing plate in place. Pry off the plate and remove the crankshaft from the end of the crankcase. Be sure

Courtesy Wisconsin Motor Corporation.

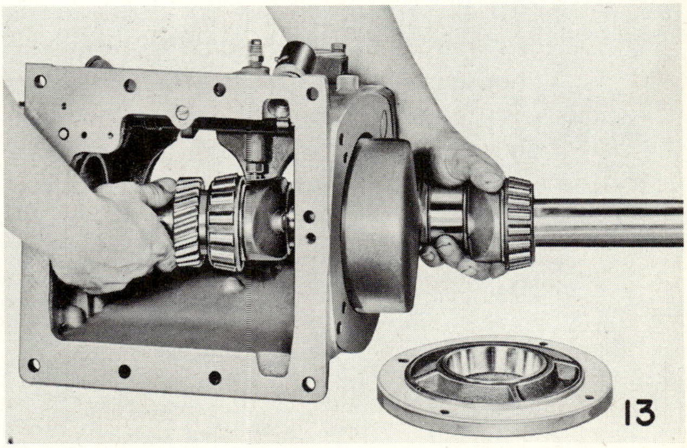
13

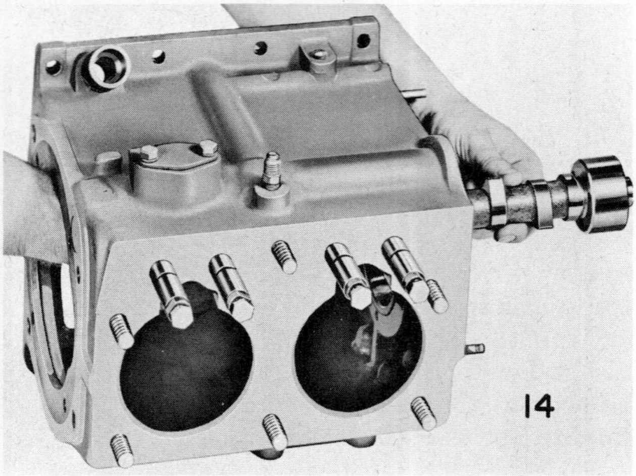

Courtesy Wisconsin Motor Corporation.

to keep the gaskets on the main bearing plate in place since they are necessary to give the proper end play to the crankshaft. This end play should be 0.001 to 0.004 inch when the engine is cold.

14. To remove the camshaft, lift the valve tappets so as to clear the cam noses as the camshaft is withdrawn. The valve tappets can be removed after the camshaft is out.

15. Clean all parts and inspect for wear. Replace any part that is worn. Use new gaskets to avoid leaks. Assembly is the reverse of disassembly.

INSPECTION OF PARTS

To check rings, insert them one at a time into the cylinder bore. Push them down with the piston to make sure they are square with the bore. Check the end gap with a feeler gauge. If the gap is 0.026 inch or more, and the cylinder bore is standard, the ring should be replaced. The gap of new rings should be 0.007 to 0.017 inch.

If the piston shows no visible signs of wear, scoring or scuffing, and the cylinder is not to be rebored, the top ring groove should be checked. To do this, clean the carbon and place a new ring in the top ring groove. Check the space with a feeler gauge. If a 0.005

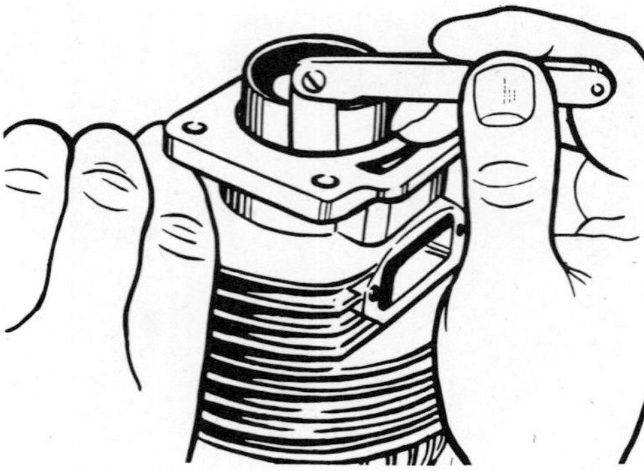

Checking the fit of a piston. Courtesy Lawn-Boy Division, Outboard Marine Corporation.

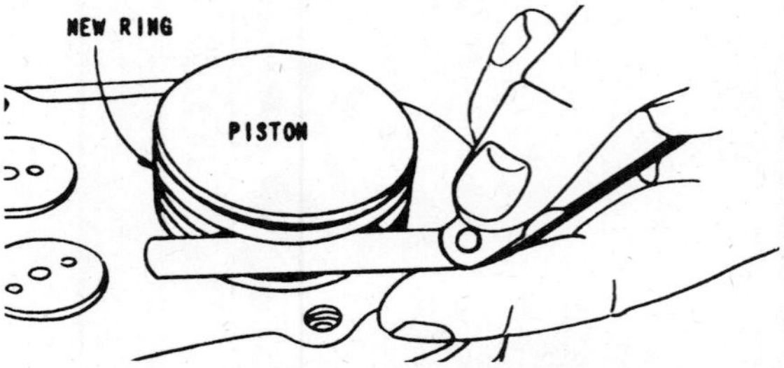

Checking the fit of a ring in the top groove to determine if excessive wear exists. Courtesy Briggs & Stratton Corporation.

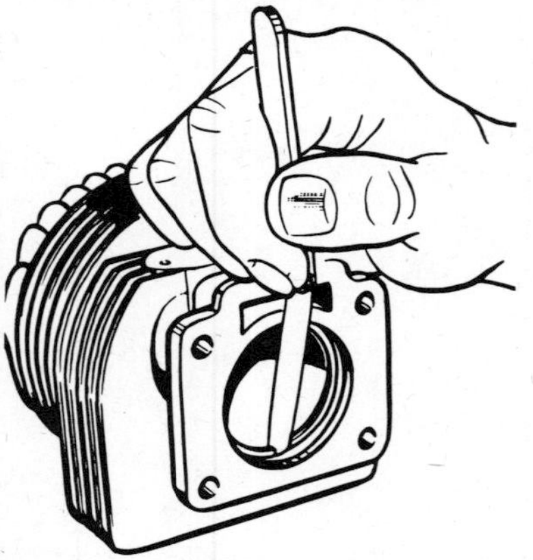

Checking the end gap of a ring. Courtesy Lawn-Boy Division, Outboard Marine Corporation.

inch feeler can be inserted on a 2-inch bore and a 0.007 inch feeler on larger pistons, the piston is worn and should be replaced.

To check the piston pin, use an accurate micrometer and take readings at several points to find the point of greatest wear. A pin 0.0005 inch out-of-round should be rejected.

If the piston pin holes in the rod and/or piston are worn, a 0.005 inch oversize piston pin is available for service. Steel connecting rods have a piston pin bushing which can be removed, a new one pressed in place, and reamed to size.

To check the connecting rod, inspect the bearing. If the crankpin bearing is scored, the rod must be replaced. Scoring is usually caused by dirty or insufficient oil. The desired clearance between rod bearing and crankpin is 0.0015 inch but should not exceed 0.0045 inch.

Do not attempt to repair the crankpin bearing in a connecting rod. If it is worn or scored the rod should be replaced.

Cylinder Inspection

Always inspect the cylinder after the engine has been disassembled. Visual inspection will show if there are any cracks, stripped bolt holes, broken fins, or if the cylinder wall is scored. Use an accurate micrometer or dial indicator to determine the size of the cylinder bore. Measure at right angles at about the center of ring travel.

Checking the crankshaft for defects. Courtesy Briggs & Stratton Corporation.

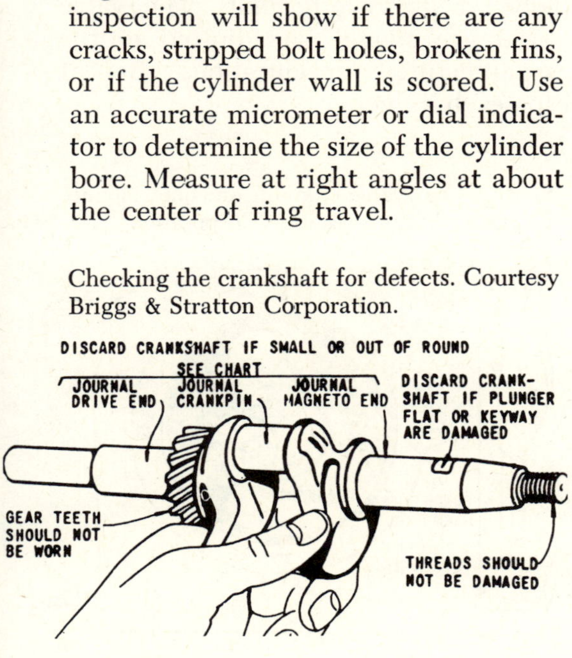

Checking the oil pump for output. Courtesy Wisconsin Motor Corporation.

If the cylinder bore is more than 0.003 inch oversize or more than 0.0015 inch out-of-round, it must be resized.

Cylinder Resizing

Always resize to exactly 0.010, 0.020, or 0.030 inch over standard size so that stock oversize rings and pistons will fit perfectly, and proper clearances will be established.

The cylinder can be resized by reboring or by honing. When using a

Using a micrometer to measure the wear in the cylinder. Courtesy Briggs & Stratton Corporation.

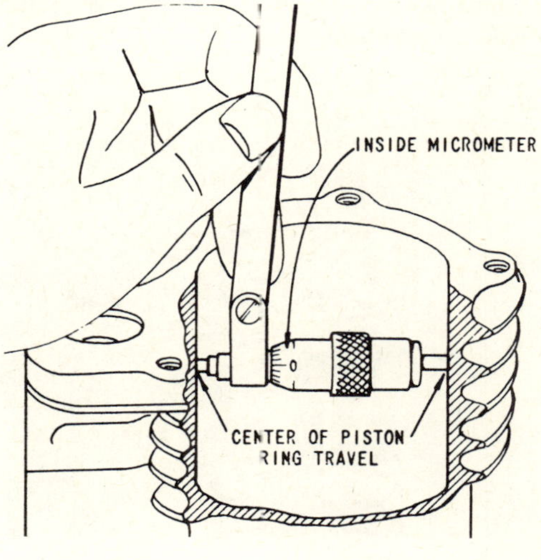

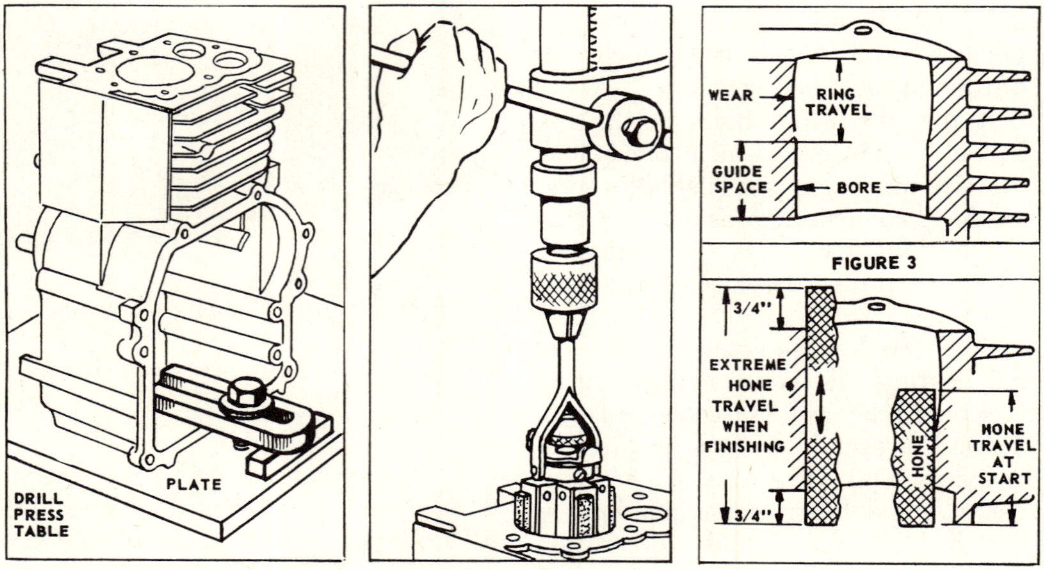

Procedure for honing a worn cylinder. Courtesy Briggs & Stratton Corporation.

hone, it is very important to start working at the bottom (the straight section) and gradually work up.

MAKING MECHANICAL ADJUSTMENTS

Adjusting Tappets

To check the tappet clearance in a four-stroke cycle engine, turn the flywheel until a valve opens to its highest point. Then turn the flywheel one complete revolution which turns the cam gear a half revolution and places the lobe at its lowest point. The valve clearance should be adjusted to the manufacturer's specifications. Repeat for each valve.

Adjusting the tappets. Courtesy Wisconsin Motor Corporation.

Setting the valve timing on one engine. Courtesy Briggs & Stratton Corporation.

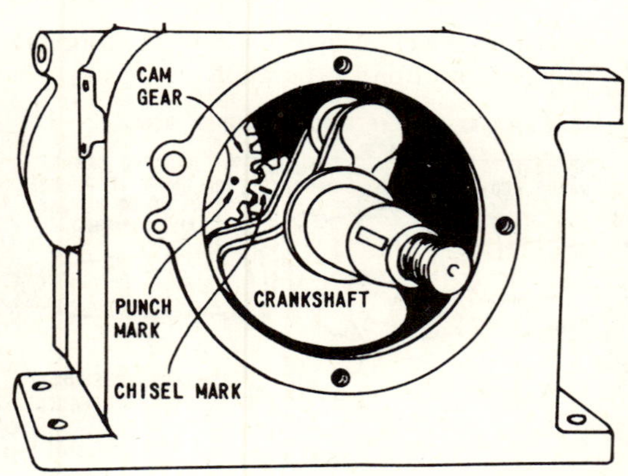

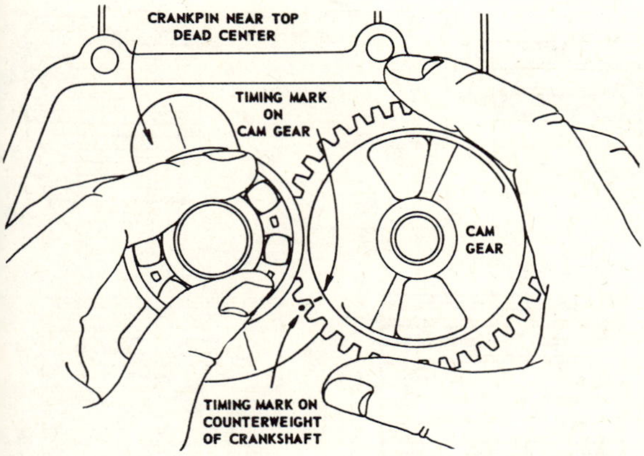

Setting the valve timing in a different gear arrangement. Courtesy Briggs & Stratton Corporation.

Correcting Crankshaft End Play

End play should be between 0.002 and 0.008 inch. Extra gaskets can be installed to increase end play. At least one gasket must be used to prevent oil leaks. If the end play is too large, a spacer washer should be placed between the crankcase and crankshaft.

Testing and correcting a connecting rod for bend. Courtesy Briggs & Stratton Corporation.

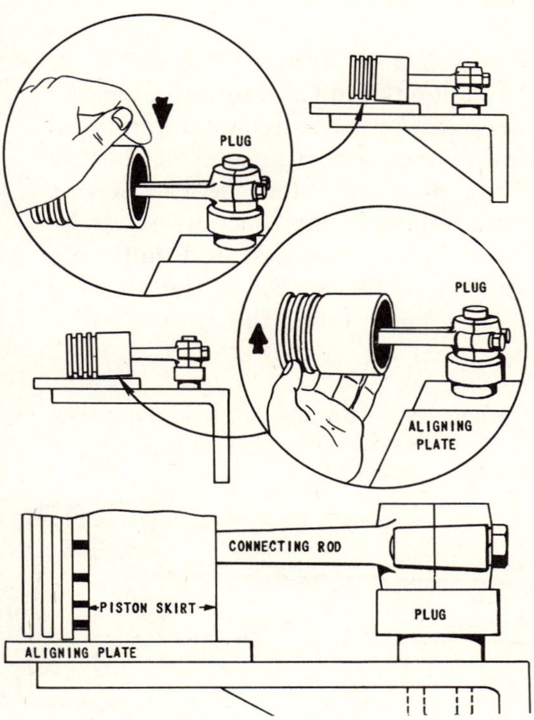

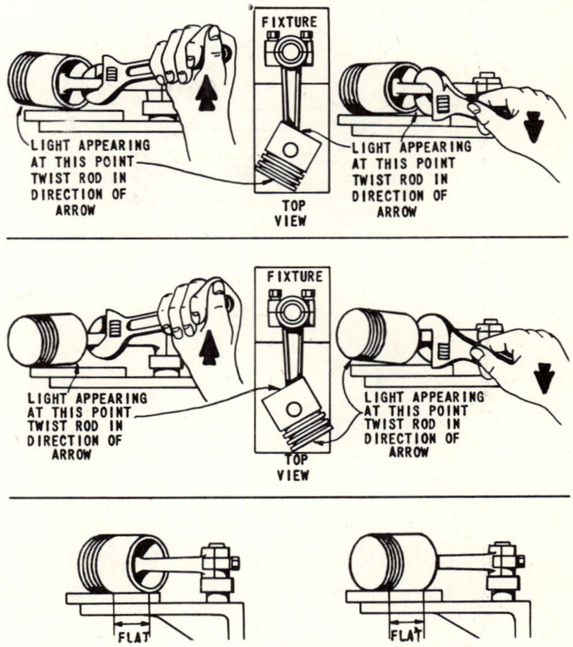

Testing and correcting a connecting rod for twist. Courtesy Briggs & Stratton Corporation.

Using a piston ring installing tool to avoid stretching the ring which would distort it. Courtesy Lawn-Boy Division, Outboard Marine Corporation.

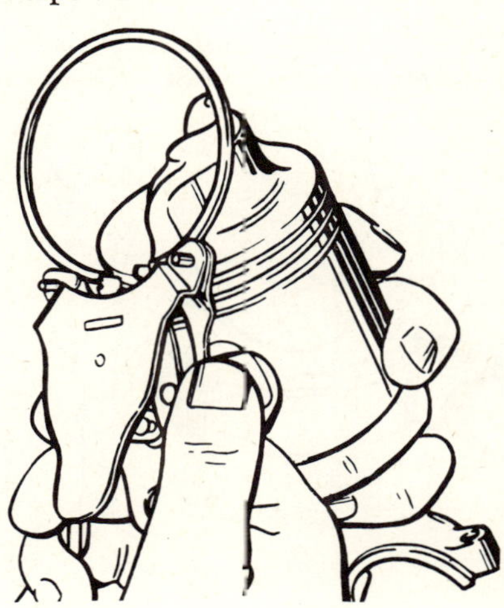

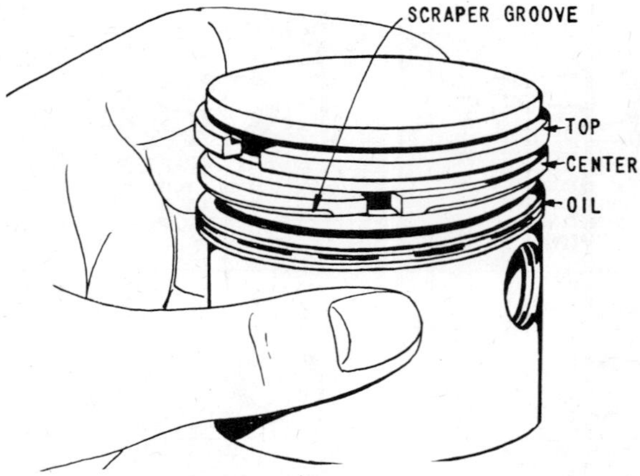

Installing the piston rings. Courtesy Briggs & Stratton Corporation.

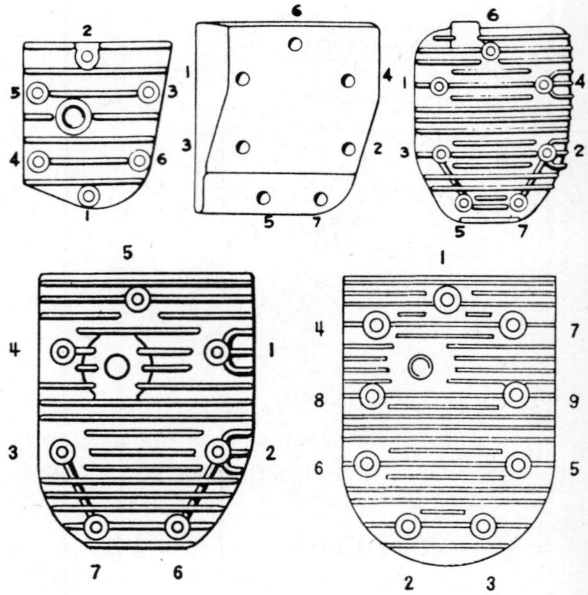
Cylinder head bolt tightening sequences. Courtesy Briggs & Stratton Corporation.

Then the gaskets should be selected which provides the correct end play.

Lapping Gasket Surfaces

Gasket surfaces tend to warp because of the thin material. Dress all such surfaces on a plate of glass covered by a sheet of No. 00 emery cloth. Move the surface you are dressing back and forth and in a figure 8. Lift the part to observe the action. If the part is warped, bright spots will show where the grinding is occurring. Continue until the surface shows a dull luster all over.

CONNECTING ROD ALIGNMENT

To align the rod and piston assembly, select the proper alignment plug and assemble the crankpin end of the rod to the plug. Tighten the cap screws so that the rod fits the plug securely. Adjust the piston and rod until the piston is in the center of the aligning plate. Note how the piston skirt

Using a compressor to install the piston assembly. Courtesy Briggs & Stratton Corporation.

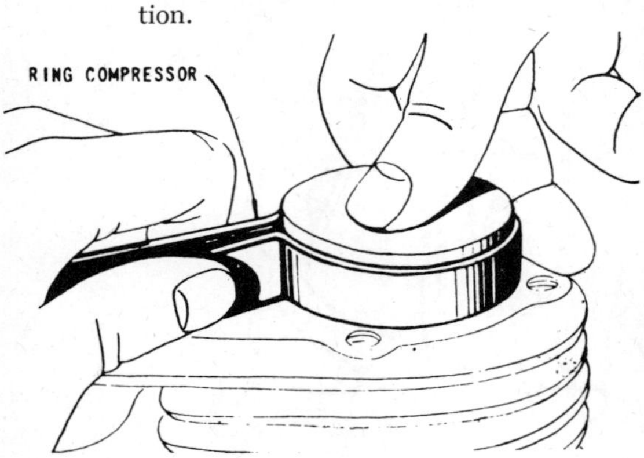

Matching marks for alignment during assembly. Courtesy Lawn-Boy Division, Outboard Marine Corporation.

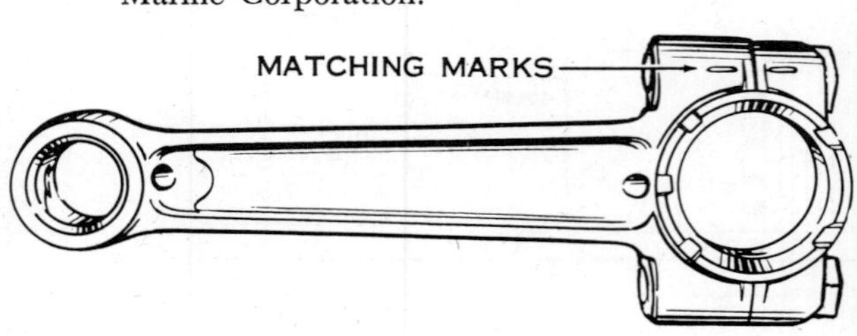

CONNECTING ROD ALIGNMENT

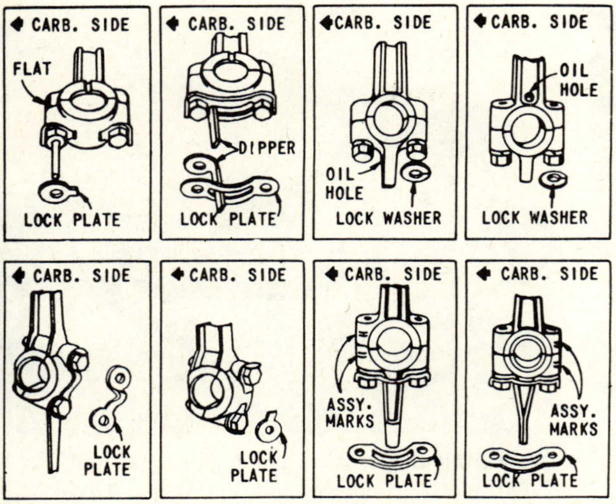

Correct assembly of the various types of connecting rods. Courtesy Briggs & Stratton Corporation.

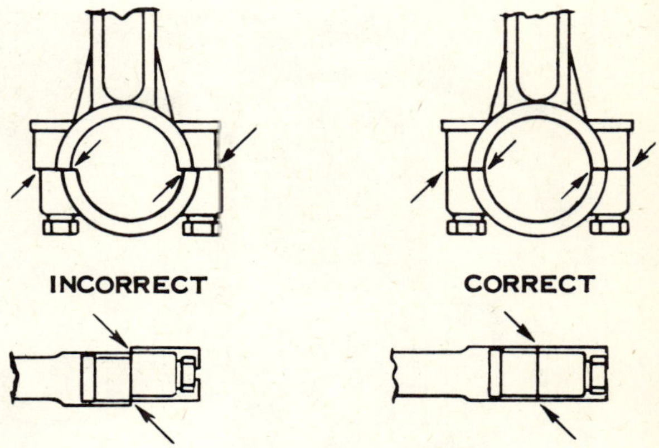

Small engine connecting rod alignment must be made by driving the two halves into alignment with a hammer after partially tightening the bolts. Courtesy Lawn-Boy Division, Outboard Marine Corporation.

touches the plate. Reverse the rod on the alignment plug to check the opposite side of the piston skirt. If the rod and piston are square, the piston skirt will touch the alignment plate evenly on each side. If it is not square, bend the rod until it is.

To test for twist, place the piston in position on the alignment plate and see if the piston touches the plate evenly when turned to either end of the piston pin travel. If the skirt does not touch evenly, swing the piston away from the plate and twist the rod with an adjustable wrench until it does.

ALIGNING THE CONNECTING ROD AND CAP

Observe the matching marks on the connecting rod and cap when assembling the parts. Draw a pencil point over the joint between the cap and rod. The joint must feel smooth. If the joint can be felt, align the cap and rod by driving the high side down with a punch and hammer. Draw up the screws tightly.

MAKING IGNITION ADJUSTMENTS

To Service the Spark Plug

Clean the spark plug with a sandblaster. Be sure to remove all abrasive with air pressure. File the inner side of the electrodes flat with a point file. If the electrodes are burned, or if the porcelain is cracked, replace the plug

Bend the ground electrode to make the spark plug gap adjustment. Courtesy Wisconsin Motor Corporation.

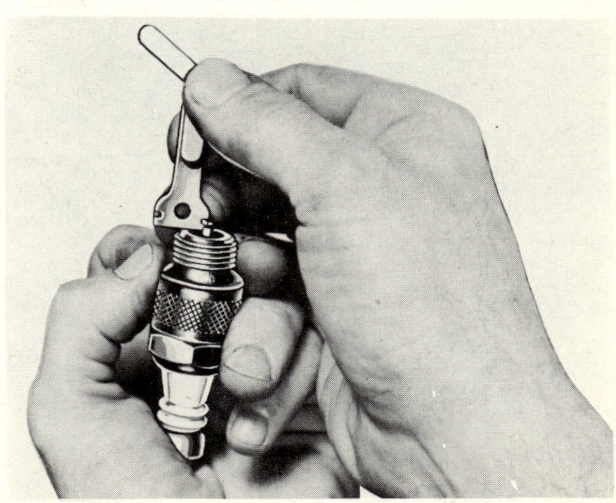

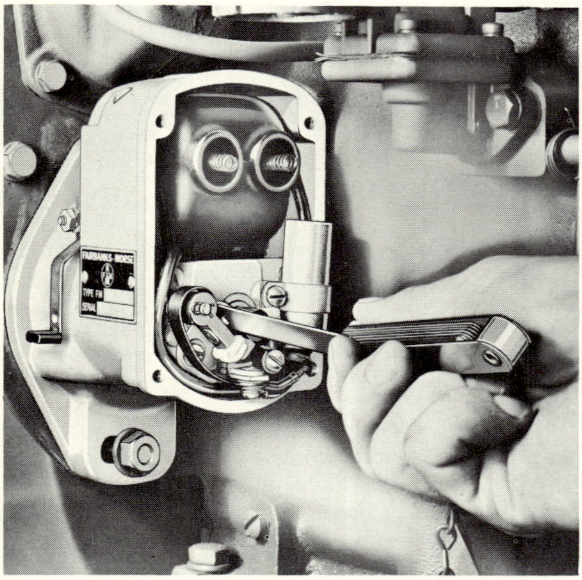

Use a cleaned feeler gauge blade to measure the point gap in a magneto. Courtesy Wisconsin Motor Corporation.

with a new one of the proper heat range. Adjust the gap to 0.025 inch. Place a little graphite on the threads to prevent sticking before installing the plug. This is especially important in an air-cooled engine.

Measuring the breaker point gap in a flywheel-type magneto. Courtesy Briggs & Stratton Corporation.

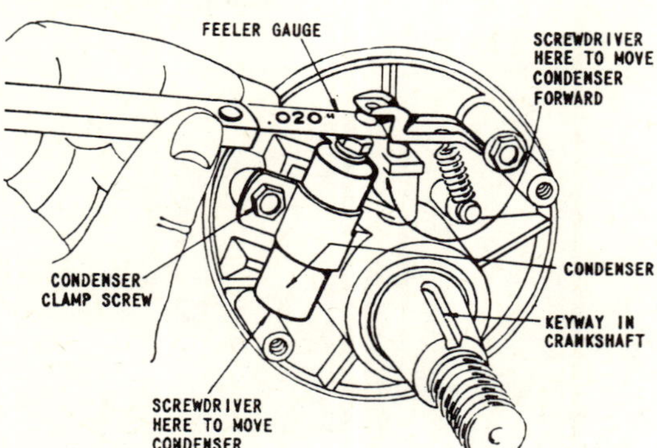

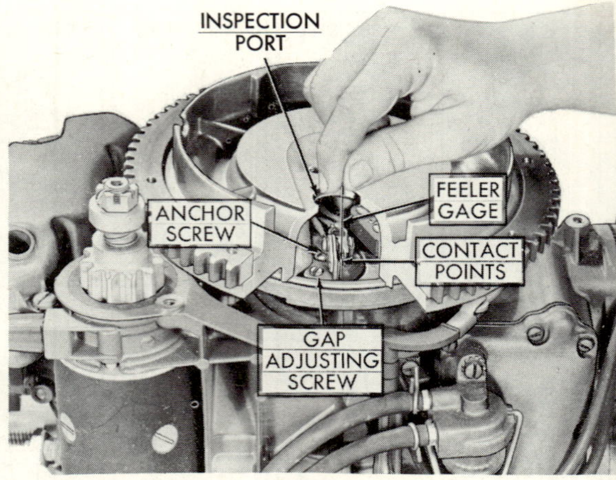

Measuring the breaker point gap in an outboard motor. Courtesy Johnson Motors Division, Outboard Marine Corporation.

To Adjust the Ignition Point Gap

The breaker point gap specification must be looked up for each engine. To make the points accessible, remove the end cover on the magneto. Rotate the crankshaft until the breaker points are wide open. The gap should be measured with a *cleaned* feeler gauge blade.

To adjust the gap, loosen the lock screw on the stationary plate, and then insert the end of a screwdriver into the

A shim is used to adjust the spacing between the armature and magnet during assembly. Courtesy Briggs & Stratton Corporation.

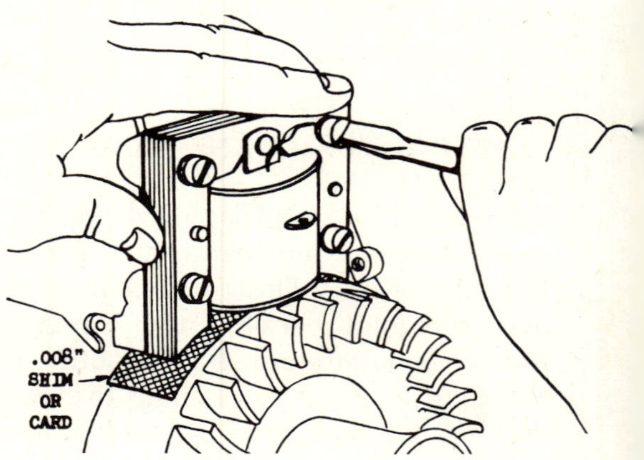

adjusting slot. Move the stationary point until the proper gap is obtained. Retighten the lock screw and then measure the gap again as it may change from the tension of the lock screw.

To Adjust the Magneto Armature Air Gap

Chalk the edges of the armature lamination and mount the flywheel in place. Turn the flywheel several revolutions by hand. Remove it and examine the edges of the laminations. The high spots will have the chalk marks rubbed off. Loosen the mounting screws slightly and tap the high side lightly to center the armature. Then retighten the mounting screws.

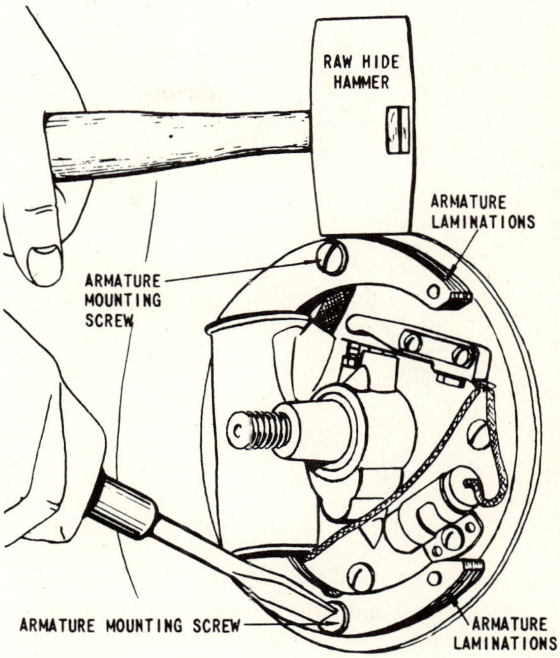

Adjusting the position of the armature in a flywheel-type magneto. Courtesy Briggs & Stratton Corporation.

TO SET THE IGNITION TIMING

(Point gap must have been set first.)

Each engine has a slightly different method of timing the ignition. The simplest method is used on Lawn-Boy engines which merely require loosening the magneto retaining plate set screw while the engine is running at its governed speed and adjusting the magneto position until the engine runs smoothest.

The Briggs & Stratton engine requires a gauge which must be mounted on the crankshaft. Place a 0.003 inch feeler gauge blade between the contact points and turn the crankshaft clockwise until the points close. Continue turning until the points *just* open and release the 0.003 inch gauge. If the gauge pointer lines up with the end of the armature shoe as shown, the timing is correct. If the pointer is past the armature shoe, the timing is late and should be corrected by installing a new fiber tip on the breaker arm. If the pointer is ahead of the end of the armature, file the bevel of the fiber tip until the pointer lines up with the end of the shoe.

The Lambretta engine is timed by

Using a pointer-type gauge to check the timing of a flywheel-type magneto. Courtesy Briggs & Stratton Corporation.

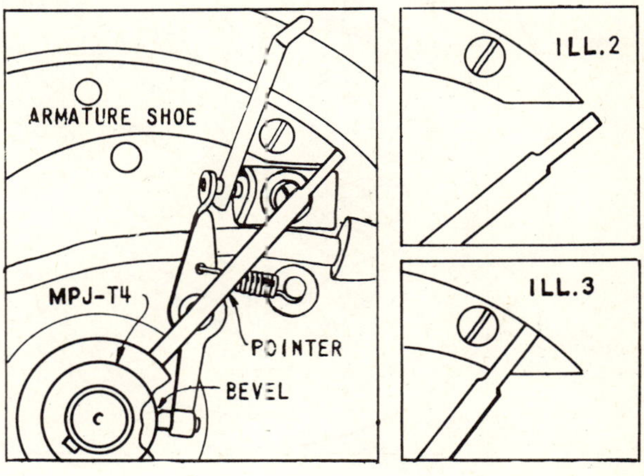

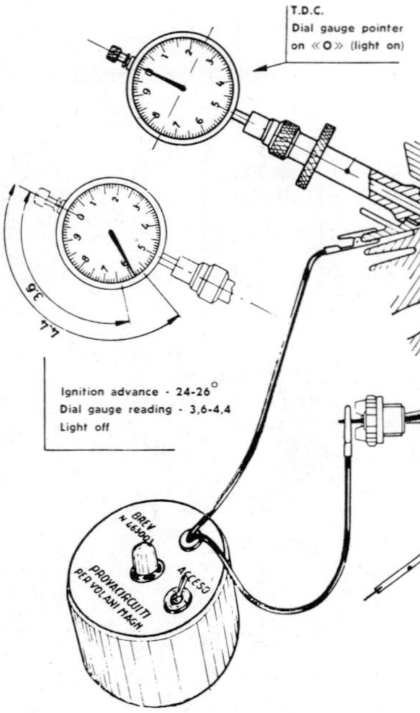

Checking the timing with a timing gauge. Courtesy Lambretta Division, Innocenti Corporation.

using a special gauge which measures piston movement in millimeters. Thread the dial gauge in place of the spark plug. Rotate the flywheel by hand until TDC is reached. Rotate the dial face until the pointer is at *zero*. Connect a circuit tester between the primary lead and ground. Rotate the flywheel until the bulb of the circuit tester goes out. The pointer movement must be between 3.6 mm (0.142 inch) and 4.4 mm (0.173 inch). If the timing is not between specified limits, loosen the stator plate and rotate it clockwise if the reading is less than 3.6 mm and counterclockwise if more than 4.4 mm. Retest and tighten the stator plate when the timing is correct.

To time a magneto with an impulse coupling, as is used on the Wisconsin

Checking the magneto timing on a Wisconsin engine. Courtesy Wisconsin Motor Corporation

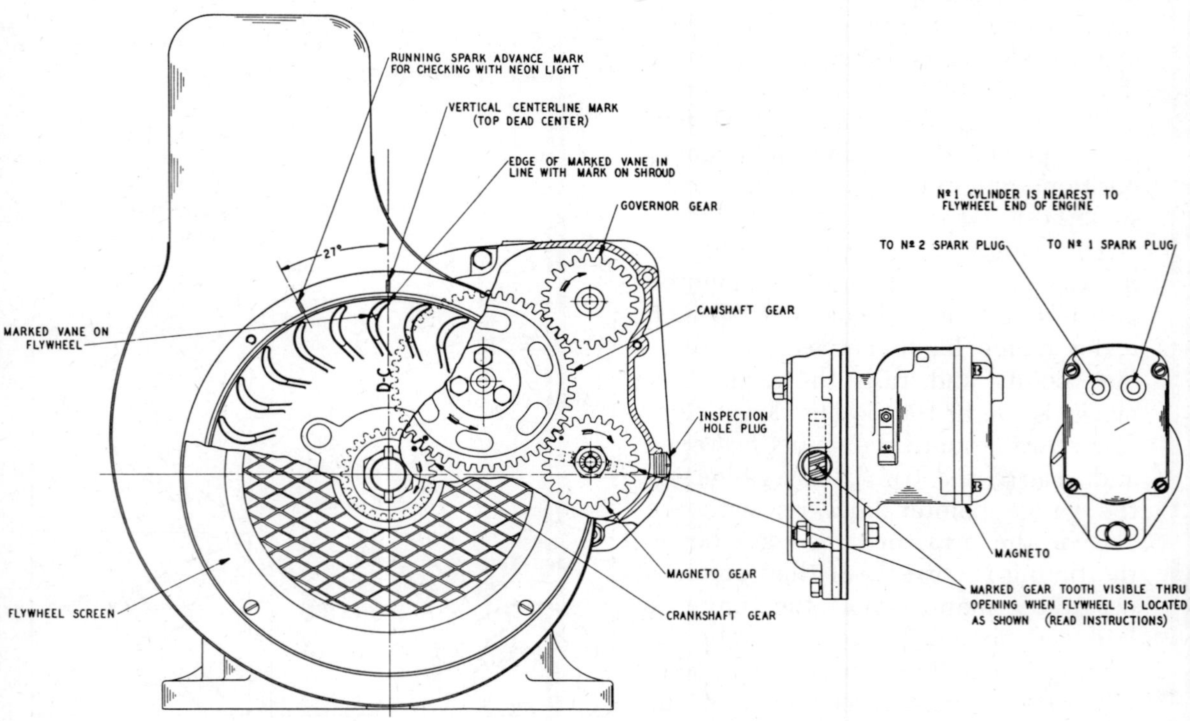

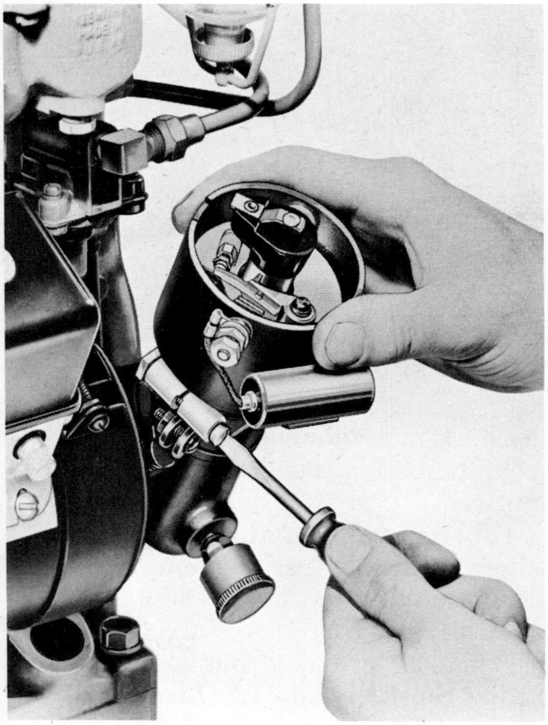

Adjusting the ignition timing on a battery-ignition system. Courtesy Wisconsin Motor Corporation.

engine, first expose the timing marks on the flywheel shroud as shown. Then remove the spark plug closest to the flywheel end of the engine. Hold your thumb over the spark plug hole while turning the engine by hand until a definite pressure can be felt, which is an indication that No. 1 piston is coming up on the compression stroke. The flywheel is marked with the letters "DC" near one of the air circulating vanes and this vane has an "X" cast into the end. Continue turning the engine until the "X"-marked vane is in line with the vertical centerline on the shroud. No. 1 piston is now at TDC.

Remove the inspection hole plug from the magneto timing opening in the engine gear cover. Turn the magneto gear in a clockwise rotation until the impulse snaps. Then, holding the gear in this position, mount magneto to the engine, meshing the gears so that the "X"-marked tooth on the magneto gear is centrally located in the inspection hole of the gear cover as shown. Tighten the magneto mounting screws. Be sure that the magneto flange gasket is in place.

The split-coil magneto, furnished on these engines, has no distributor and, therefore, provides two sparks simultaneously with each rotation. One spark is used for ignition while the other passes harmlessly through the exhaust gases. Therefore, either tower can be connected to No. 1 spark plug. The proper spark advance is 27° at operating speed. To check this, connect a neon timing light to No. 1 spark plug and the "X"-marked vane should line up with the 27° mark located about 2½ inches to the left of the vertical centerline when the engine is operating at normal speed.

SERVICING THE CARBURETOR

Because most small engine fuel tanks are above the carburetor, close the shut-off valve first. Disconnect the fuel line at the carburetor, disconnect the throttle linkage, remove the air cleaner, and take out the two screws holding the carburetor to the intake manifold.

To check the throttle shaft and bushings for wear, place a short piece of stock on the upper carburetor body just below the throttle shaft as shown. Press the shaft down firmly and measure the distance between the shaft and stock with a feeler gauge. Pull the shaft up and again measure the dis-

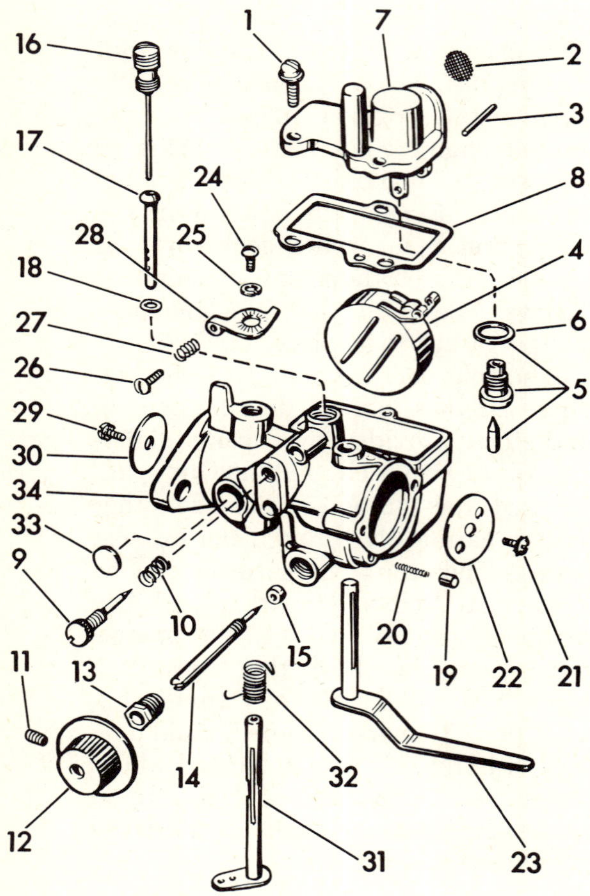

Exploded view of a typical small-engine carburetor. Courtesy Lawn-Boy Division, Outboard Marine Corporation.

Checking the throttle shaft for wear. Courtesy Briggs & Stratton Corporation.

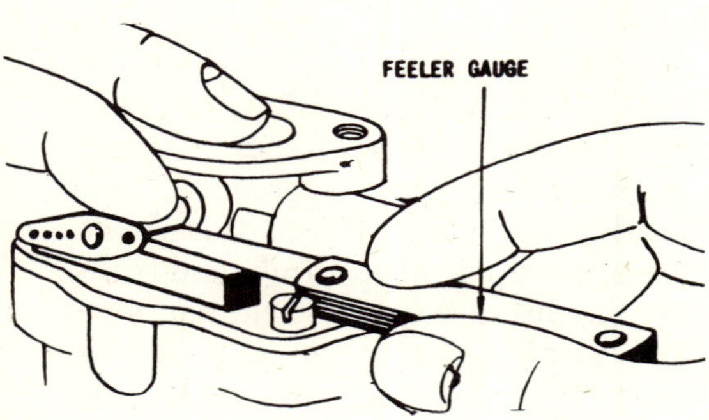

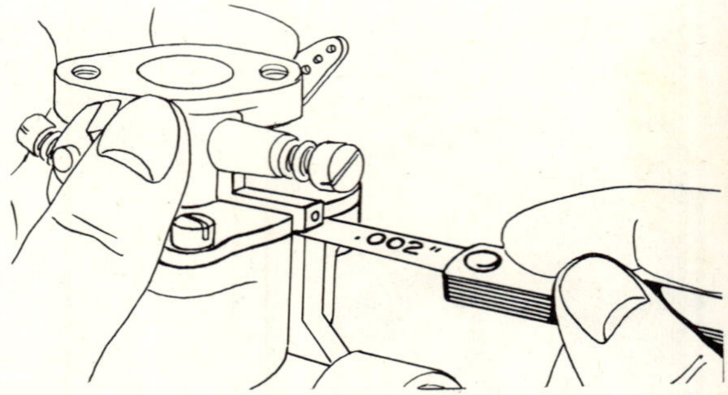

Checking the throttle body for warpage. Courtesy Briggs & Stratton Corporation.

tance. If the difference exceeds 0.010 inch, the bushings and shaft should be replaced.

To disassemble the carburetor, loosen the needle packing and remove the needle valve. Use a thin, blunt screwdriver to remove the nozzle which must be taken out before the bodies can be separated. Remove the four screws holding the bodies together. Remove the idle valve adjusting screws.

Remove the pin which holds the float in place. Take out the inlet

Removing the main nozzle. Courtesy Briggs & Stratton Corporation.

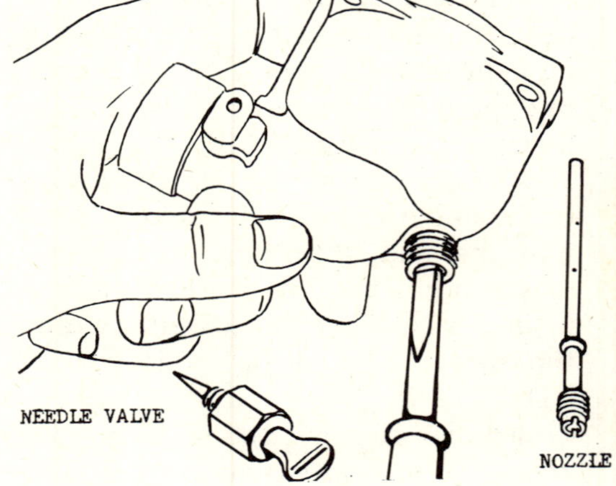

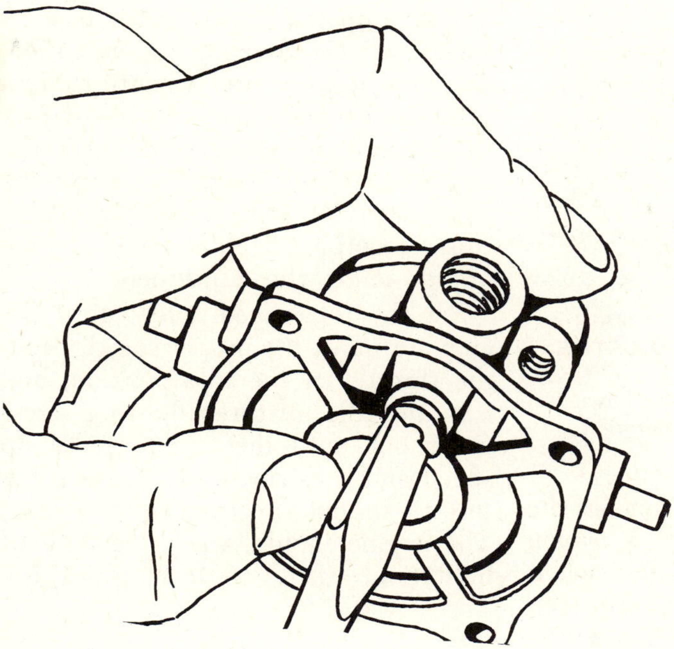

Removing the needle-valve seat. Courtesy Briggs & Stratton Corporation.

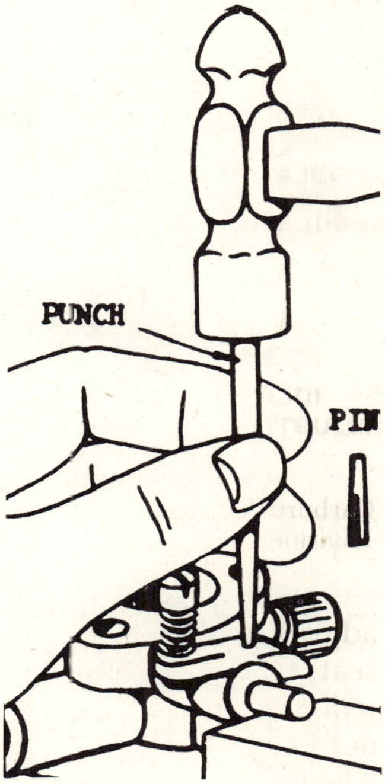

Removing the pin locking the stop arm to the throttle shaft. Courtesy Briggs & Stratton Corporation.

needle. If the float contains gasoline, it should be replaced. Use a wide screwdriver to remove the inlet valve seat. Pull the venturi from the lower body.

Remove the throttle valve. Use a thin punch to drive out the pin which retains the throttle stop to the shaft, and then pull out the shaft. If the shaft and bushings are worn, they should be replaced. Some carburetors have no bushings and the upper casting must be replaced when worn.

To pull the throttle shaft bushings, use a ¼ inch x 20 tap held in a vise. Screw the bushing onto the tap. Press new bushings into the casting with the vise jaws. Run a 7/32 inch drill or reamer through the bushings so that the shaft turns freely.

When reassembling the carburetor, use new gaskets, a new needle valve and seat assembly, and new parts where necessary. Be sure to set the float to specifications. Turn in the idle

Using an E-Z Out to remove the throttle shaft bushings. Courtesy Briggs & Stratton Corporation.

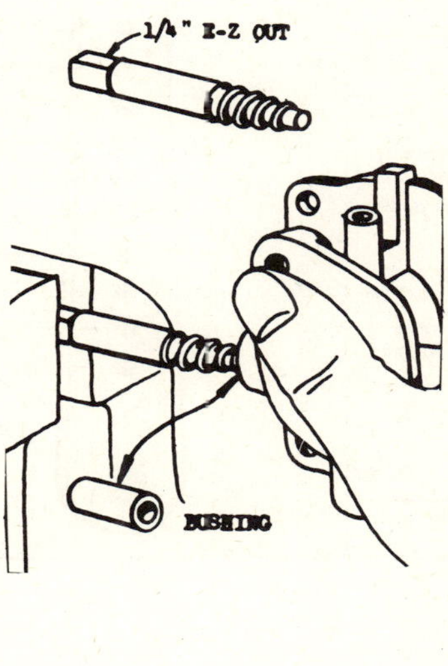

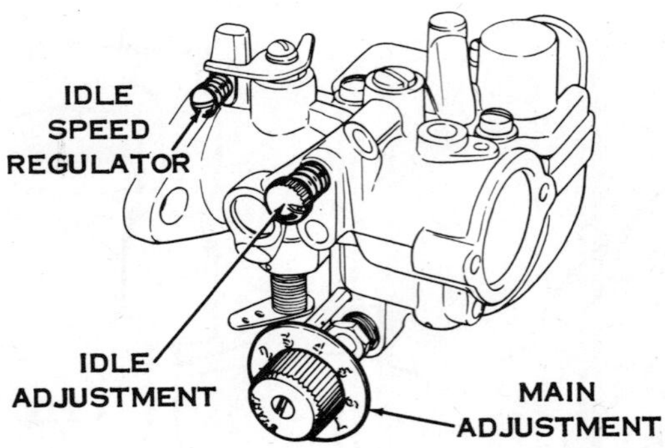

Carburetor adjustments. Courtesy Lawn-Boy Division, Outboard Marine Corporation.

adjusting screw until it just touches the seat. **CAUTION:** *Don't force it or you will groove the needle.* Now back it out ¾ turn for a preliminary adjustment. The final adjustment will be made with the engine running.

Governor hole positions and resulting engine rpm's. Courtesy Wisconsin Motor Corporation.

LOAD R.P.M.	NO LOAD R.P.M.	HOLE NO.
1400	1550	2
1500	1650	3
1600	1725	3
1700	1850	4
1800	1925	4
1900	2025	5
2000	2150	6
2100	2225	6
2200	2350	7
2300	2425	7
2400	2550	8
2500	2650	9
2600	2725	9

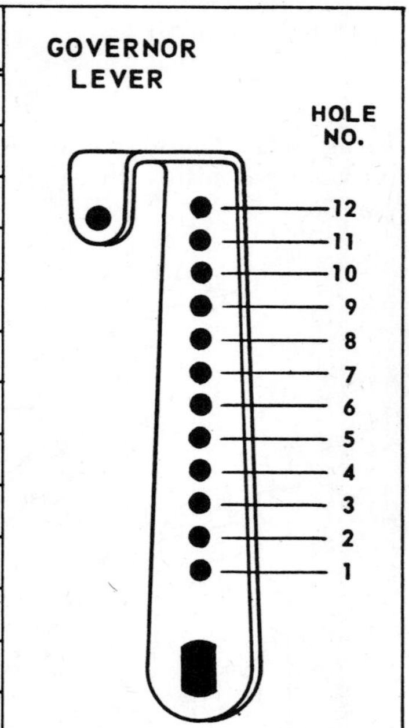

Warm up the engine and adjust its speed to the governed specification. Turn the needle valve in until the engine misses, and then turn it out past the smooth operating point until the engine runs unevenly. Then turn the valve in just enough so that the engine runs smoothly. Tighten the packing nut to retain the adjustment.

Set the idle speed screw until the engine runs at the proper speed. Hold the throttle in the idle position, and turn the idle mixture adjusting screw in or out until the engine picks up speed and runs smoothly. Release the throttle so that the governor takes over. The engine should accelerate without hesitation. If it sputters, adjust the

Checking the operation of the governor. Courtesy Wisconsin Motor Corporation.

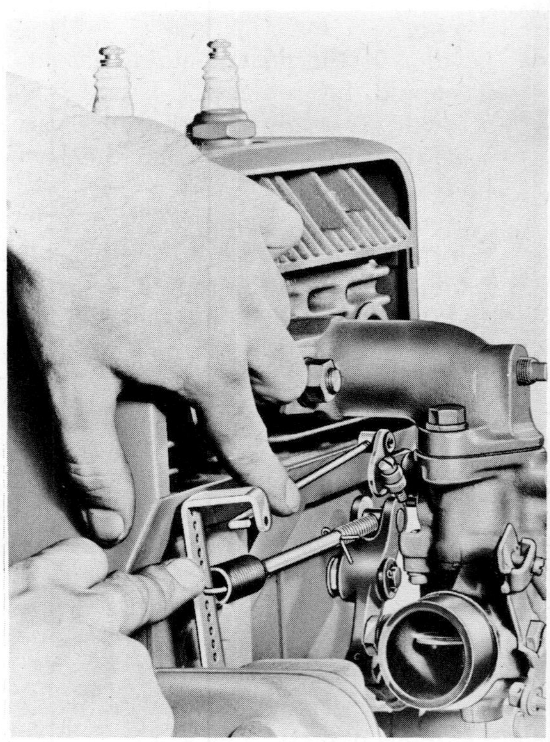

Governor adjustments on one type of Briggs & Stratton engine. Courtesy Briggs & Stratton Corporation.

Governor adjustments on another type of small engine. Courtesy Briggs & Stratton Corporation.

high-speed needle valve to a slightly richer position.

SERVICING THE GOVERNOR

A governor is used on many small engines to keep the engine speed constant under varying loads. It consists of a set of flyweights which bear against a flanged sliding sleeve, moving it back and forth as the flyweights move in or out. The motion of the sleeve is transmitted to the governor lever, which in turn is connected to the carburetor throttle lever. A spring, connected to the governor lever, tends to hold the governor flyweights to their inner position and also to hold the carburetor throttle wide open. As the engine speed increases, centrifugal force of the flyweights acts against the spring and closes the throttle to

Operation of the governor mechanism. As the engine speeds up, the weights (arrows) fly out, causing the control arm to close the carburetor throttle. Courtesy Lawn-Boy Division, Outboard Marine Corporation.

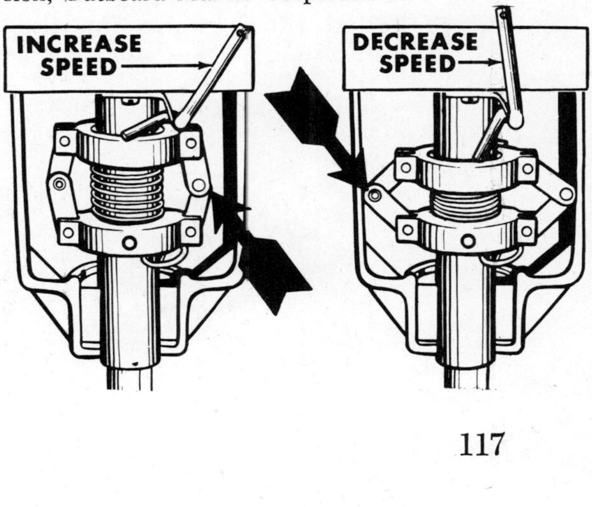

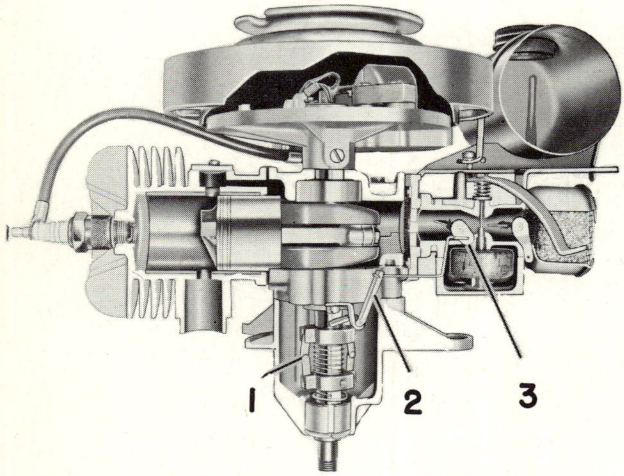

Parts of the governor. 1. Weights; 2. Operating arm; 3. Carburetor throttle arm. Courtesy Lawn-Boy Division, Outboard Marine Corporation.

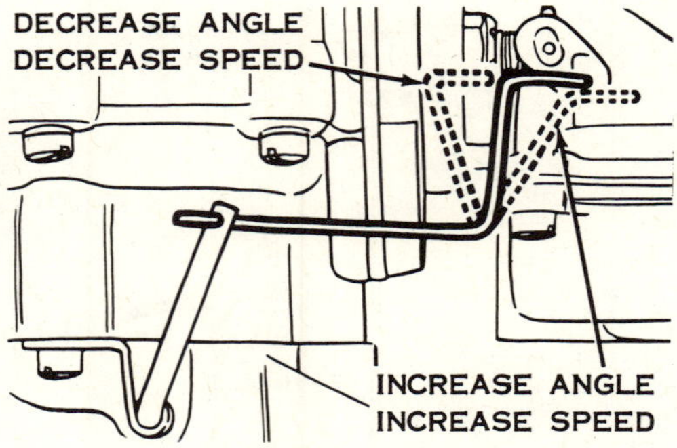

Bending the operating arm results in governed speed changes. Courtesy Lawn-Boy Division, Outboard Marine Corporation.

a point where engine speed is kept constant under varying load conditions. This speed can be varied by adjusting the governor spring position.

The control rod from the governor to the carburetor must be adjusted to the proper length otherwise the governor action will be faulty. With the engine stopped, the governor spring must hold the flyweights in, and the control rod must be of such length as to hold the carburetor throttle wide open. The accuracy of this adjustment can be tested by disconnecting the rod from the governor lever, and then pushing the rod toward the carburetor as far as it will go which opens the throttle wide. The bent end of the rod should then exactly register with the hole in the governor lever. If it does not, the rod should be screwed into or out of the swivel block on the carburetor lever until it registers.

SERVICING STARTERS

Starters are either of the kick type, hand crank, or rope pull. Electric starters are serviced the same as the automotive types. The kick starters and rope pull types require spring replacement and the latter sometimes requires replacement of the rope.

To make the necessary repairs, disassemble the starter cover and spring

Position of the rope-type starter. Courtesy Briggs & Stratton Corporation.

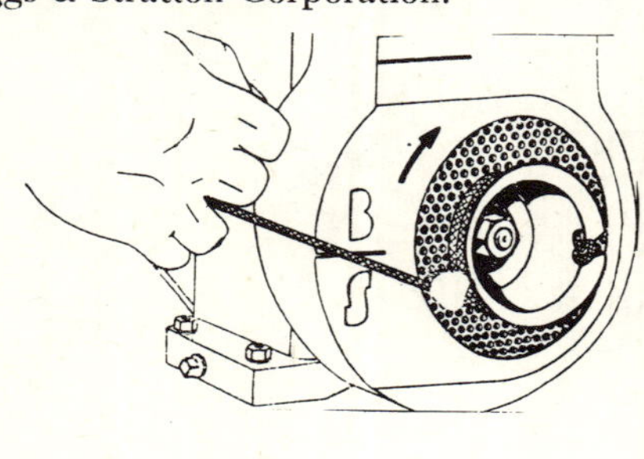

Pedal-type starter adjustments. Courtesy Briggs & Stratton Corporation.

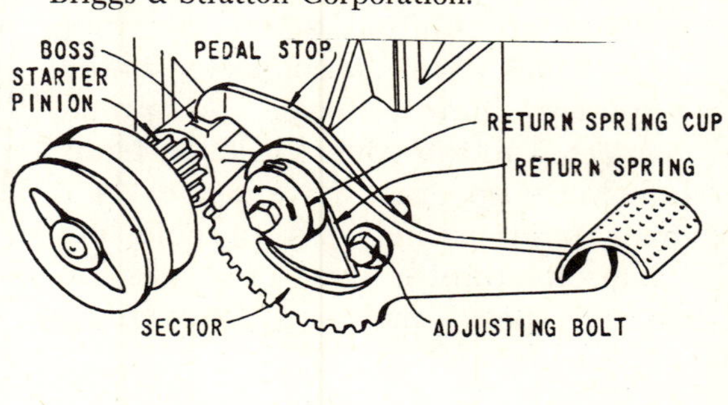

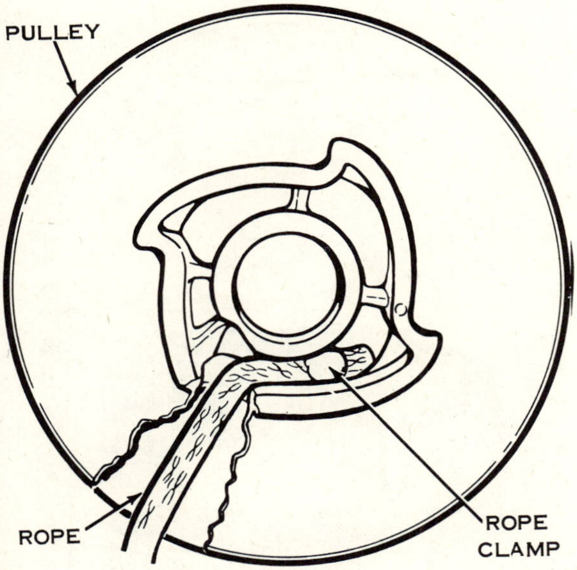

Burning the end of a nylon rope swells it for a tight fit. Courtesy Lawn-Boy Division, Outboard Marine Corporation.

housing. Singe the end of the new rope with flame to stiffen it. Place the spring in position with its loop over the pulley lug. Assemble the pulley over the starter cap with the spring running through the slot. Securely tighten the cap to the pulley. Wind the rope

Installing the rope. Courtesy Lawn-Boy Division, Outboard Marine Corporation.

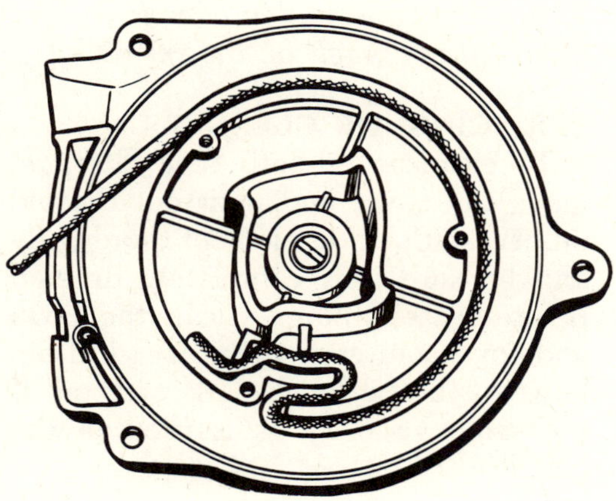

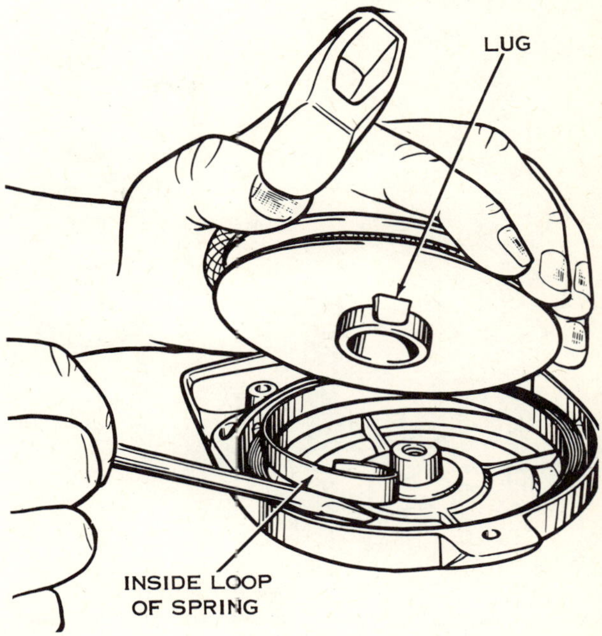

Correct installation of the starter spring. Courtesy Lawn-Boy Division, Outboard Marine Corporation.

Pulling on the rope to tension the spring. Courtesy Lawn-Boy Division, Outboard Marine Corporation.

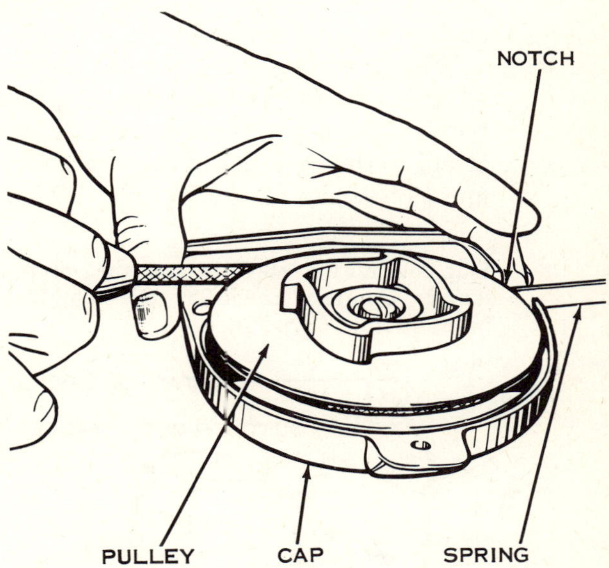

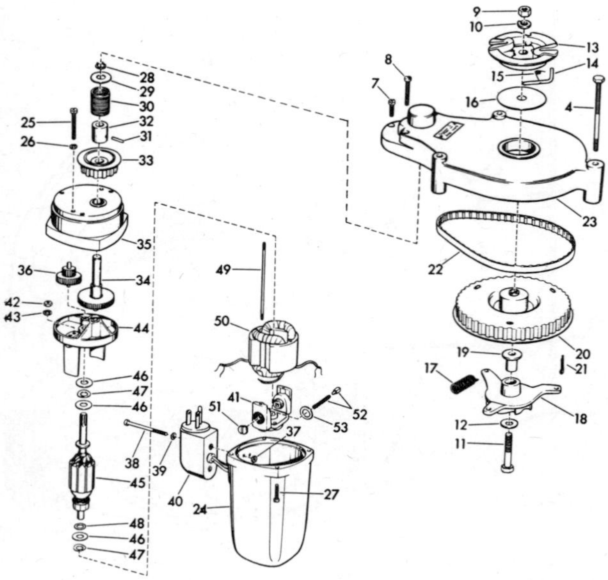

Exploded view of an electric starter. Courtesy Lawn-Boy Di ision, Outboard Marine Corporation.

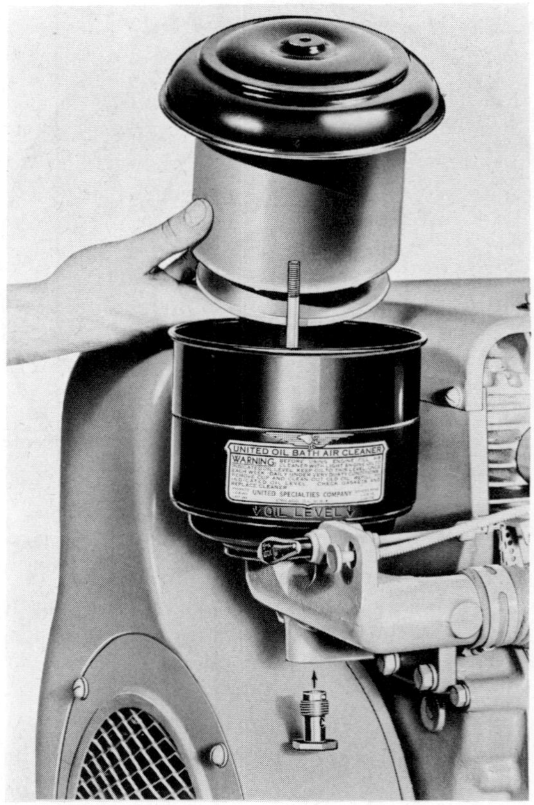

Servicing the air cleaner. The fuel filter has also been removed (bottom) for servicing. Courtesy Wisconsin Motor Corporation.

around the pulley. Then pull it to draw the spring into the cap. Repeat this operation several times until all of the spring is drawn into the cap. The bent end of the spring will catch on the anchor lug of the cap.

With the rope fully wound on the pulley, turn the pulley against spring tension until the rope handle is adjacent to the rope slot in the cap. There must be some tension on the spring, otherwise the handle will not return all the way when released. **CAUTION:** *Never wind the pulley over two turns after tension is felt on the spring.*

Plugging in the cord to use the electric starter. Courtesy Lawn-Boy Division, Outboard Marine Corporation.

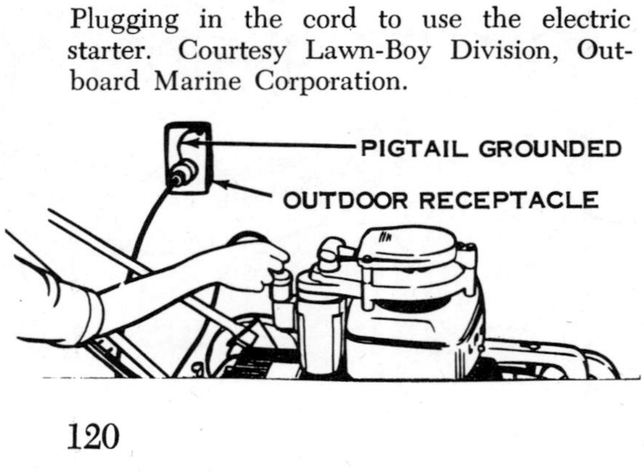

SERVICING AIR CLEANERS

To clean an oil-bath type filter, remove the cover and pour out the old oil. Wash the filter element thoroughly and be sure it is completely drained before reassembling. Clean the bowl and cover and wipe them dry. Fill the bowl to the oil level mark with oil of the same viscosity as is used in the crankcase.

To clean the felt type filter, rinse it in gasoline. Blow out all dirt from the element. If the element is gummy, it should be replaced.

To clean the moss-type filter, rinse it in gasoline. Fill the element with oil through the holes. Let the excess oil drain. If the moss is gummy, replace the element.

CHECK YOUR KNOWLEDGE

1. What are some of the uses to which small engines are put?
2. What operating difference is sometimes found in small engines?
3. What two main differences are there between repairing small and large gasoline engines?
4. What is the main difference between small- and large-engine carburetors?
5. What device is used to keep small-engine speed constant?
6. How does a small-engine ignition system differ from that used on a large engine?
7. What kind of fuel is best suited to small engines?
8. What kind of gasoline is recommended for small engines of the two-stroke cycle type?
9. What type of oil is best suited to small engines of the two-stroke cycle type?
10. What fuel-oil mixture is recommended for two-stroke cycle engines?
11. Make a list of the troubles which are specifically found only in two-stroke cycle gasoline engines.
12. What advantages are there in using several pans to hold the small parts when disassembling a small engine?

Self-check tests

TRUE-FALSE

1. Small engines are used mainly for transportation.
2. All small engines operate on the two-stroke cycle principle.
3. A small engine requires no valves or camshaft.
4. The fuel and oil must be mixed for a four-stroke cycle engine.
5. Many small engines have governors.
6. A small-engine carburetor contains the same number of jets as does an automotive carburetor.
7. The magneto-ignition system requires no storage battery.
8. The magneto generates its own primary current.
9. High octane fuel should be used in small engines to minimize spark plug fouling.
10. White gasoline forms gum when the engine is storaged.

MUTIPLE CHOICE

1. Which of the following is *not* a type of starter for small engines: a) kick; b) impulse; c) electric; d) rope?
2. Which of the following is *not* a part of a magneto: a) coil; b) condenser; c) battery?

COMPLETION

1. A ________ is mechanically connected to the carburetor butterfly valve in order to maintain the desired engine speed.
2. The magneto generates its own ________ current.
3. The best fuel to use in a two-stroke cycle engine is ________ gasoline.

MATCHING

1. Two-stroke cycle engine	1. Controls the carburetor throttle valve
2. Four-stroke cycle engine	2. Ignition
3. Governor	3. Contains a camshaft
4. Magneto	4. Contains ports

CHAPTER 4
GENERAL SAFETY PRACTICES

ATTITUDES

The practice of safety in the shop goes beyond the knowledge of the proper use of hand tools and equipment. A most important consideration is your attitude. You must understand the hazards of the job you are about to undertake and must appreciate the need for applying such safety practices that will protect you from injury.

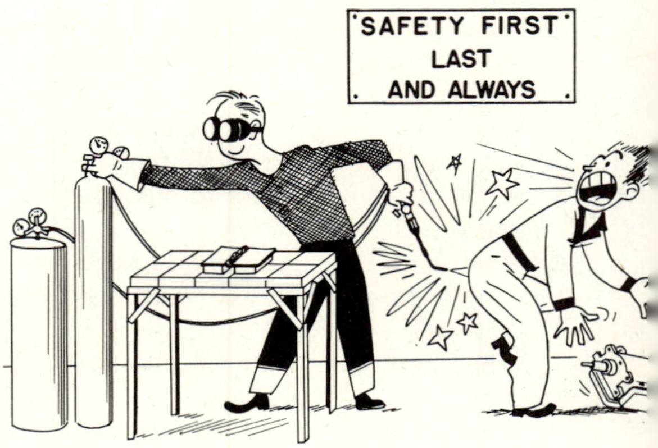

Carelessness can result in hurt "feelings."

Look before you leap. Courtesy Los Angeles Times, Home Magazine.

Awareness of personal safety comes only through an understanding of the dangers which are present. Most accidents are caused by thoughtlessness, so you must be on guard at all times.

It is most important to give the job your undivided attention. "Visiting" is not permitted because it distracts you and may lead to an accident.

Horseplay, scuffling, punching, or playing pranks is dangerous. Some boys cannot pass by without striking a classmate, and the immediate response is to strike back. In no time at all both boys are scuffling. This childish action may result in a fall or possible injury

You don't have to be crazy to take chances, but it helps! Courtesy National Safety Council.

from sharp tools, steel benches, and heavy equipment which are always present in a shop. Keep such physical activity for the athletic field where it is an approved activity.

Walk—avoid running! A running person cannot always keep from slipping with the possibility of serious injury. Bumping the operator of a machine might cause him to have an accident which would be your fault. Cultivate personal caution at all times.

A cautious person is one who knows and observes safe procedures. Learn that caution and foresight pay off. A person who has learned to cultivate personal caution has trained himself to visualize the results of his actions.

Make it a habit to remove any article lying on the floor before someone trips over it. Learn to recognize unsafe conditions. In doing so you are protecting yourself and fellow workers. • Get into the habit of removing sharp pointed tools from bench edges where you, or someone else going by, may get hurt—this is an application of personal caution. • Store sharp-edged tools in racks and not in drawers where someone could cut himself when picking up other implements.

• When stacking stock, make sure that it is arranged with the large boxes on the bottom of the pile.

• Round stock should be secured with chocks to prevent the pile from rolling.

• Keep stock out of aisles or passageways because someone may fall over it.

• A protruding nail is an invitation to

There's always a safe way for every job. Courtesy National Safety Council.

a first-aid station; either hammer it down or pull it out. • Remove any splinters from boxes or pieces of wood which could puncture your skin. • Scraps of sharp metal on the floor could pierce someone's shoe, causing a serious wound—pick them up and place them in the scrap box.

• Wipe spilled oil or grease off the floor—*even though you did not spill it.* Preventable falls send far too many people to the hospital.

Let's become safety conscious!

Always expect the unexpected. Courtesy National Safety Council.

PERSONAL SAFETY

• Report any injury to your instructor immediately. No matter how small the injury—report it! It is far better to waste antiseptic than to ignore one small wound which may later become infected. There must never be a letdown in the battle against infection.

• If something gets into your eye, it is dangerous to rub it. Instead report to your instructor immediately. You may press the particle into the eyeball and scratch its surface. Or, if you use an unsterile cloth, you may cause serious eye infection and possible blindness.

• Be sure to wash your hands after handling caustics, acids, or batteries to avoid getting chemicals on your skin and into your eyes. • Wash your hands thoroughly before eating and always after using the lavatory. The need for personal cleanliness cannot be overstressed. Many skin conditions can be traced to improper cleanup after work. A good soap, thoroughly applied, removes dirt and grime as well as acids and caustics.

• Asking for help when lifting a heavy object is a sign of mental strength, not weakness. Many serious injuries are caused by lifting improperly, lifting objects too heavy for one man, or lifting unwieldy shapes. More than 25% of all disabilities are caused by the improper handling of materials. This is the largest single cause of disability from all accidents. Sprains, strains, and hernias—the results of improper lifting—are painful and disabling.

• When lifting a heavy object, place your feet close to the object for proper balance. Keep your elbows as straight

Get help for heavy or awkward loads. Courtesy National Safety Council.

as possible and bend the knees while gripping. Use your large leg muscles to lift—not the back muscles. Keep your back straight. When lifting a heavy object with the help of others, be sure that a signal is given by one of the team so that excess strain is not placed on any one member of the group. Teamwork accomplishes much more than individual effort.

• Carry all objects in such a manner that you can see clearly where you are going. Long objects should be carried by two people to protect others. If a long object has to be carried by one person, make sure that the dangerous front end is held low so it can be seen. The rear end will be high enough to pass over people's heads as you make turns.

• Power tools are provided with guards to prevent accidents. They are placed there for your protection. Be sure to call the instructor's attention to any loose or missing guard.

PROTECTIVE CLOTHING

Wearing the proper clothing is very important when working in the shop. Each shop requires a specialized protective garment depending on the nature of the work. Some shopwork demands additional temporary protection when doing special jobs. This may take the form of goggles and gloves when welding; goggles or face shield for grinding; or a rubber mat when working around live electrical circuits.

In general, coveralls are necessary in the auto shop to protect you from dirt and grime. White coveralls are the ones most commonly used in the trade. Dark ones are sometimes worn because they do not show the dirt; however, dirty coveralls of any color stain car upholstery and are not pleasant to look at. It is good practice to develop a habit of cleanliness by changing your coveralls at least once a week.

You can always use a world of protection. Courtesy National Safety Council.

Gloves should be worn when handling hot objects, especially when welding. They are also good protection when working with shattered glass in automotive work. **CAUTION:** *They must not be used when buffing or working around machinery as there is danger of the gloves being caught and pulling your hand into the machine.*

Safety shoes with extra strong toe

When the tool's not right, the guy's not bright. Courtesy National Safety Council.

Tool misuse is tool abuse. Courtesy General Motors Corporation.

supports protect your feet from falling objects. They are also good protection against burns when welding and are especially desirable for the protection afforded the ankles against bruises and scuffs when working under a car.

SAFETY IN THE USE OF HAND TOOLS

The hand tool is still very important, despite the development of complex machine tools. Everyone has need for hand tools, but not everyone knows how to use or care for them properly. A recent study of shop accidents showed that 66% of all injuries in one year were caused by misuse of the common hand tool. Most hand tool accidents are caused by (1) improper storage, (2) failure to keep the tool in good condition, (3) using the tool improperly, and (4) failure to use the right tool for the job.

Hand tools should be stored in a tool rack which has a place for each tool; they should not be stored in drawers or boxes. Tools should be cleaned frequently and stored in a dry place to prevent rusting. A light coating of oil will keep them bright under adverse conditions. Knowledge gained in this section of the book will assist you in learning how to use each hand tool properly so that undue strain will not be placed on it, causing it to slip—which could result in injury—and how to select the tool which is designed for the job at hand so that it will not be subjected to unnecessary stress.

PEEN HAMMERS

The *peen hammer* is a very commonly used tool in power mechanics. The flat end of the head is called the *face*, and the other end the *peen*. If the

You can't always make a hit by flying off the handle. Courtesy General Motors Corporation.

peen is ball shaped, it is known as a *ball-peen hammer*. The hole for the handle is called the *eye*.

There are right and wrong ways to hold a hammer. The beginner usually "chokes" the handle by gripping it too near the head. When you want to strike a heavy blow, grip the handle near the end. This increases the length of the *lever arm* and makes the blow more effective.

It is dangerous to use a hammer with a loose head, as it may fly off. The eye of the hammer head is tapered both ways from the center. After the handle is inserted, a steel wedge is driven into the end to expand the handle and hold it firmly. If the wedge starts to loosen, drive it in again; if it is missing, replace it.

When you use a hammer on a machined surface, protect the surface with a piece of soft brass or wood. Special soft hammers of plastic and lead faces are frequently used to prevent damaging finely machined surfaces. If the work is hardened steel, there is additional danger from flying particles. A hard steel hammer face, striking a hard steel object, is likely to fracture. The fractured piece may fly off with the speed of a bullet and can create a serious puncture wound.

Using a hammer handle for bumping purposes is likely to split it which can cause splinter wounds.

SCREWDRIVERS

Every mechanic uses a screwdriver. Most screwdriver handles are made of plastic or wood. The steel portion ex-

A screwdriver is not meant to be used as a pry. Courtesy General Motors Corporation.

Hold it! Get a chisel before you wreck that screwdriver. Courtesy General Motors Corporation.

Oh, come now! Let's use the proper-sized tool. Courtesy General Motors Corporation.

tending from the handle is called the *shank*, and the flattened end which fits into the slot of the screw is called the *blade*.

A screwdriver is designed to withstand considerable twisting force but not to be used as a pry or chisel. The blade is hardened to keep it from wearing; it may snap if too much bending strain is applied.

A screwdriver tip must fit the slot properly; otherwise, it will slip. Courtesy General Motors Corporation.

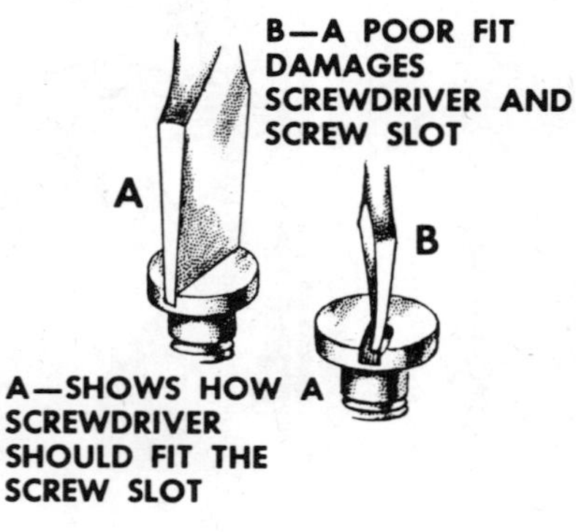

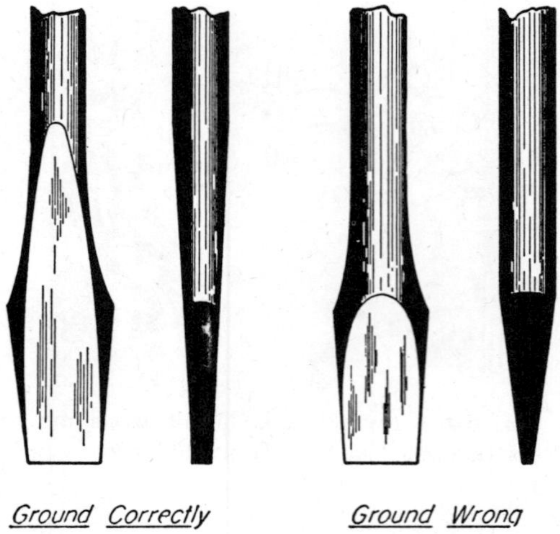

A correctly ground screwdriver blade has its sides parallel. Courtesy Federal-Mogul Service.

Using a screwdriver as a cold chisel will split the handle. However, most screwdrivers made for automotive work have steel shanks all the way through the handles so that you can tap on them if necessary. *To remove a rusted screw*, tap the end of a steel-shank

Easy does it! Courtesy General Motors Corporation.

screwdriver while holding it at an angle, in order to clean out the screw slot. After the slot is cleaned, holding the screwdriver vertically and tapping on the handle will seat it in the slot and help to jar rusted threads loose.

Not enough emphasis can be placed on selecting the *correct-sized screwdriver* so that the thickness of the blade fits the screw slot, which keeps the slot from burring and reduces the force required to keep the blade in the slot. A screwdriver which slips is dangerous. Clamp the work in a vise while removing screws; otherwise, the screwdriver may slip and puncture your palm, which is very painful and makes a wound difficult to treat.

The tip of a *correctly ground screwdriver blade* should have the sides almost parallel. When dressing the blade, grind it so that the faces taper in very slightly for a short distance back of the tip so that it will stay down in the slot. A blade with too much taper raises out of the slot and may slip off the screw head.

A screwdriver in a leg pocket of your coveralls should be carried *point down,* as bad scratches are possible if it is carried point up. The exposed blade can also mar the finish of a car when you lean over the fender. Sharp-edged tools should be carried in a covered sheath, not in your pocket, as stab wounds may result.

WRENCHES

Wrenches are designed to be inserted from the side of a nut (as in the case of an open-end, adjustable, monkey, or pipe wrench), or over the top (socket or box wrench) to hold it firmly for removing or tightening.

If you have to pull hard on a wrench, make sure that it *seats squarely* or it may slip. Pushing on a wrench is dangerous; if the nut breaks loose suddenly you may skin your knuckles. However, if you do have to push on the wrench, use the palm of your hand.

Pushing on a wrench is the wrong approach to this nutty problem. Courtesy General Motors Corporation.

Box wrenches are very popular because of their secure grip. The wrenches are usually made with twelve notches arranged in a circle, and they are, therefore, called twelve-point wrenches. As they cannot spread on a

Always pull so that the strain is against the solid jaw of an adjustable wrench to keep it from slipping. Courtesy General Motors Corporation.

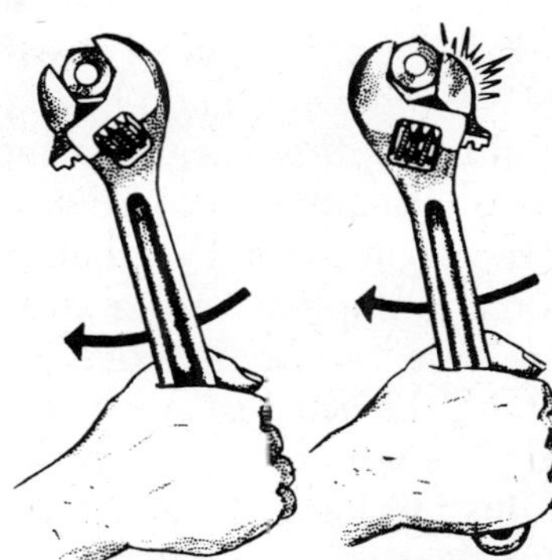

Mushroomed chisel—
Flying missile.
Ere you start upon your mission,
Have your tools in good condition.

nut, there is little danger of slippage.

Adjustable wrenches are shaped somewhat like an open-end wrench but have one adjustable jaw. The angle of the opening is 22½ degrees. Although adjustable wrenches are especially convenient at times, they are not intended to take the place of standard open-end, box, or socket wrenches. Their grip is not as secure, and they are more likely to slip. Smaller adjustable wrenches are principally used when loosening an odd-sized nut.

Adjustable wrenches are not intended for hard service; treat them gently. Whenever you do have to exert any real force, there are two important points to remember: (1) Always place the wrench on the nut so that the pulling force is applied to the strong, stationary side of the handle, as it can withstand the greatest stress. (2) After placing the wrench on the nut, tighten the adjusting knurl so that the jaws fit the nut securely, to prevent slipping.

Hammering on a wrench or extending it with a pipe places an excessive strain on a wrench which it is not designed to take.

CHISELS

Chisels are used for cutting metal. The one most commonly used is the flat cold chisel to split nuts, cut rivets, cut heads off rusty screws, and remove bushings. The width of the cutting edge denotes its size. The average mechanic's tool kit contains three chisels of ¼, ½, and ¾ inch sizes.

Always use a chisel that is *big enough for the job.* Use a hammer that is heavy enough. The larger the chisel, the heavier the hammer should be. Ordinarily a chisel should be held in the left hand with the thumb and first finger about an inch from the upper end. Hold the chisel with a steady but rather relaxed grip. If you should miss the chisel and strike your hand, this tends to lessen the effect of the blow. The best practice, of course, is not to miss the chisel! *Keep your gaze on the cutting edge.* Strike one or two light blows to test your swing, and increase the force of the blows as you become more sure.

When chipping, always wear goggles to protect your eyes. If others are working close by, make sure they are protected from flying chips by a screen, or else chip in a direction which is

Mushroomed heads should be ground off chisels whenever they are sharpened. Courtesy General Motors Corporation.

clear of workers. Remember—the time to take precautions is before you start a job, not after someone is injured.

Clamp small pieces of work rigidly in a vise. Chip toward the solid or stationary jaw of the vise and away from yourself.

Each time you sharpen a chisel, inspect the other end to see if it has become upset or mushroomed, as a result of hammering. It is dangerous to use a chisel or punch which has been upset because fragments may fly off and cause injury. Grind off the extra metal so that the end of the chisel is slightly tapered and comparatively flat across the top.

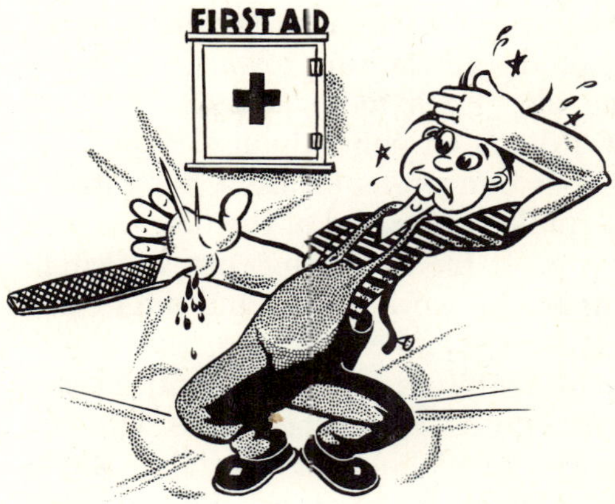

He got the point—do you? Always use a handle on a file. Courtesy General Motors Corporation.

FILES

A mechanic's tool kit cannot be considered complete without several files. The portion of the file on which the teeth are cut is called the *face,* and the tapered end which fits into the handle is called the *tang.* The *teeth,* which do the cutting, are set at an angle across the face.

A file with a single row of parallel teeth is called a *single-cut file;* those with a second row of teeth crossing the first row in a criss-cross pattern are called *double-cut files.* Criss-crossing

Types of single- and double-cut files. Courtesy General Motors Corporation.

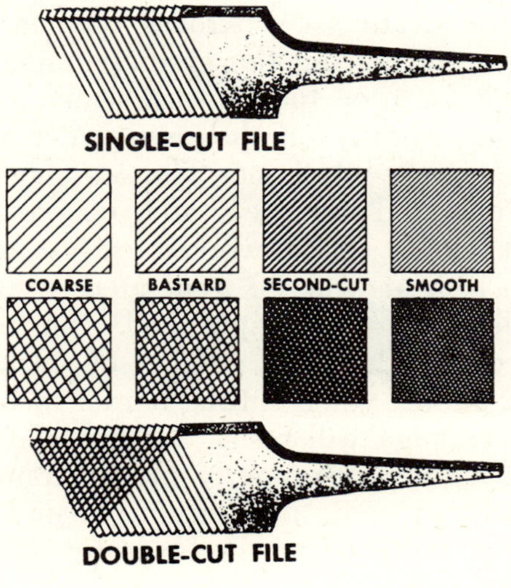

Prying with a file can ruin your profile. Courtesy General Motors Corporation.

the cuts produces a far greater number of teeth—all slanting toward the tip of the file. Each tooth acts like the end of a diamond-pointed chisel.

It is dangerous to use a file without a handle, as the end of the tang is quite sharp. If the file "hangs up" by catching on the work, your hand might jam against the end of the tang, resulting in a very painful puncture wound. *To put a handle on a file,* select one with the proper-sized hole for the tang, slip it on, and then tap it on the bench to seat it.

Be sure that the work to be filed *is held securely* because loose work permits the file to chatter, which quickly dulls the teeth. *Exert just enough pressure* to keep the file cutting. Insufficient pressure allows the file to slip over the work which rapidly dulls the teeth. *Raise the file slightly* on the return stroke in order to clear the work. This does not apply in filing soft metals because drawing the file back over the work aids in cleaning the teeth.

Use a *file brush* frequently to remove the chips from between the teeth. Tapping the end of a file against the vise will dull or even shatter it. Keep your files *hanging in a rack* rather than lying in a drawer with other tools. Files are extremely hard and brittle, and are, therefore, easily damaged by such treatment.

Using a file for *prying* may snap it in two. *Hammering* on a file may cause it to shatter with chips flying off in all directions.

HACKSAWS

The two parts of a hacksaw are called the *frame* and *blade.* Practically all hacksaws are made with an adjustable frame designed to take blades

Take it easy—do it right.
You don't have to saw with all your might!
Courtesy General Motors Corporation.

which are 8, 10, or 12 inches long. Most frames are made with a *pistol grip.*

When placing a blade in a hacksaw frame, adjust the frame for the length of the blade, allowing sufficient threads to permit the blade to be stretched tightly. Always place the blade on the pins with the teeth pointing toward the *front* of the frame.

To use a hacksaw correctly, the following pointers must be observed: Make steady, slow strokes with even pressure, placing the greatest amount of pressure on the forward stroke and lifting slightly on the return, making each stroke a full one. The saw should be operated at a rate of 40 to 60 strokes per minute. If it is worked too fast, too much heat will be generated which will draw the temper and ruin the blade. Reduce the number of strokes per minute when cutting harder metal.

Breakage will result if the blade is not tightened properly in the frame. Breakage may also occur if undue strain is placed on the blade by twisting the frame while cutting. It is important that the blade has the correct number of teeth per inch for the metal being cut. Using the wrong blade will shorten its life.

Finer-toothed blades will cut slower but with less risk of tooth breakage. Use blades with fourteen teeth per inch for cutting 1 inch or thicker cast iron, machine steel, brass, copper, aluminum, or bronze: Use blades with eighteen teeth per inch for cutting from ¼ to 1 inch annealed tool steel, high speed steel, rail, or light structural shapes. Use blades with twenty-four teeth per inch for cutting ⅛ to ¼ inch material. Use blades with thirty-two teeth per inch for materials thinner than ⅛ inch.

CHECK YOUR KNOWLEDGE

1. Why is the proper attitude so important in the shop?
2. Why is visiting undesirable during working hours?
3. What danger is there in poking a fellow student for fun?
4. Why is it so important to avoid running in the shop?
5. What is meant by personal caution?
6. What are the seven personal caution practices listed, which can be applied to everyday shop situations?
7. Why is it so important that every injury be reported—no matter how slight?
8. What are two possible results of getting something in your eye?
9. Why should you avoid rubbing your eye if you do get a foreign particle in it?
10. Why is it so important to wash your hands after handling chemicals or before eating?
11. What are some possible results of lifting a heavy object improperly?
12. What is the proper way to stand when lifting a heavy object?
13. How should a long object be carried by one man to protect fellow students?
14. How should stock be arranged to minimize danger of its falling?
15. Why is it so important that guards are in place and working properly before operating power equipment?
16. What procedure should be followed if a guard is missing?
17. Why is it desirable to wear white coveralls?
18. Why should coveralls be changed at least once a week?
19. Why should you remove gloves when working around moving machinery?
20. What three protections are afforded by the use of safety shoes?
21. What are the four most common causes of hand tool accidents?
22. How should tools be stored to minimize hand injuries?
23. How do properly sharpened hand tools minimize cuts and bruises?
24. What is the proper way to grip a hammer handle?
25. How is the hammer head secured to the handle?
26. What protection is needed for a finished steel surface which must be hammered on?
27. What hazard is there in using a steel hammer on hard steel parts?
28. What damage may be caused by using a screwdriver as a pry?
29. What procedure is recommended for removing rusty screws?
30. Why is it so important to select the correct-sized screwdriver for the job?
31. Why should the work be clamped in a vise when removing or replacing screws?
32. How should the screwdriver blade be ground?
33. What precaution is discussed for carrying a screwdriver in your pocket?
34. Why is it dangerous to push on a wrench?
35. What advantage does a box wrench have over an open-end wrench?

36. What is the disadvantage of an adjustable wrench?
37. What two precautions are given for the safe use of adjustable wrenches?
38. Why is it inadvisable to increase the leverage of a wrench?
39. How should a chisel be held?
40. What precaution should be taken against flying chips when using a hammer and chisel?
41. Toward what part of the vise should your blows be directed when using a hammer and chisel?
42. What danger is there in using a chisel with a mushroomed head?
43. What danger is there in using a file without a handle?
44. How much pressure should be kept on the file on the forward stroke? On the return stroke?
45. How should files be stored to minimize dulling the teeth?
46. Why is it dangerous to use a file as a pry?
47. How should the chips be removed from the file teeth?
48. How should the hacksaw teeth point when properly installed?
49. What will occur if the hacksaw is worked too fast?
50. What will occur if the blade is not tightened securely in the frame?

Self-Check Tests

TRUE-FALSE

1. A skillful worker is always a safe worker.
2. If you find an article lying on the floor which may be tripped over, the safe procedure is to locate the person who dropped it.
3. Sharp-edged tools should always be stored in a drawer where you cannot accidentally cut yourself.
4. Small injuries need only be tended by stopping the blood flow.
5. One out of every eight disabilities is caused by improper lifting.
6. When lifting a heavy object, use your back muscles by bending over to grasp the object.
7. Gloves are the best protection to prevent cuts when working around machinery.
8. The best place to store hand tools is in a drawer where they can easily be found.
9. A sharp hand tool is much more dangerous than a dull one.
10. Always pull on a wrench when loosening a tight nut.
11. Adjustable wrenches are especially desirable to use because they fit all sizes of nuts and are just as safe as box wrenches.
12. When using an adjustable wrench, place it on the nut so that the pulling force is applied to the stationary jaw.
13. It is important to keep your file clean of chips by tapping it occasionally on the end of the vise.
14. A file is made of good hard steel, therefore, it makes an excellent pry.

MULTIPLE CHOICE

1. Small injuries should be: a) ignored; b) washed thoroughly; c) reported immediately.
2. One out of every: a) two; b) four; c) eight; d) sixteen disabilities is caused by improper lifting.
3. When carrying a long object, the front end should be kept: a) low; b) level; c) high.
4. a) $1/3$; b) $1/2$; c) $2/3$ of all shop injuries are caused by misuse of the common hand tool.

INDEX

D

E

F

U

V

W